▶ The Public Sector R&D Enterprise

DOI: 10.1057/9781137542090.0001

Science, Technology, and Innovation Policy

Series Editor: Albert N. Link

Managing Editor: Leila Campoli

Science, Technology, and Innovation Policy is a scholarly series for academics and policy makers. Topics of interest include, but are not limited to, the economic foundations of science, technology, and innovation policy; the impact of science, technology, and innovation policy on economic growth and development; science, technology, and innovation policy as a driver of sustainability and social well-being; and the application of methods and models for quantifying the social consequences of science, technology, and innovation policy.

Titles include:

Peter D. Linquiti
THE PUBLIC SECTOR R&D ENTERPRISE
A New Approach to Portfolio Valuation

Peter D. Blair
CONGRESS'S OWN THINK TANK
Learning from the legacy of the office of technology assessment (1972–1995)

Albert N. Link
BENDING THE ARC OF INNOVATION
Public support of r&d in small, entrepreneurial firms

palgrave▸pivot

The Public Sector R&D Enterprise: A New Approach to Portfolio Valuation

Peter D. Linquiti

George Washington University, USA

palgrave
macmillan

DOI: 10.1057/9781137542090.0001

First published in 2015 by
PALGRAVE MACMILLAN®
in the United States—a division of St. Martin's Press LLC,
175 Fifth Avenue, New York, NY 10010.

Where this book is distributed in the UK, Europe and the rest of the world,
this is by Palgrave Macmillan, a division of Macmillan Publishers Limited,
registered in England, company number 785998, of Houndmills, Basingstoke,
Hampshire RG21 6XS.

Palgrave Macmillan is the global academic imprint of the above companies
and has companies and representatives throughout the world.

Palgrave® and Macmillan® are registered trademarks in the United States,
the United Kingdom, Europe and other countries.

ISBN: 978-1-137-54210-6 EPUB
ISBN: 978-1-137-54209-0 PDF
ISBN: 978-1-137-54208-3 Hardback

Library of Congress Cataloging-in-Publication Data is available from
the Library of Congress.

A catalogue record of the book is available from the British Library.

First edition: 2015

www.palgrave.com/pivot

DOI: 10.1057/9781137542090

Contents

DOI: 10.1057/9781137542090.0001

DOI: 10.1057/9781137542090.0001

DOI: 10.1057/9781137542090.0001

List of Figures

DOI: 10.1057/9781137542090.0002

List of Tables

DOI: 10.1057/9781137542090.0003

Acknowledgments

This book has benefited substantially from the input of a number of individuals, including Nicholas Vonortas, Arun Malik, Alex Triantis, Joseph Cordes, Gregory Tassey, and anonymous referees.

In addition, comments and suggestions received at the 2010 International Real Options Conference at LUISS Guido Carli University in Rome, the 2011 Atlanta Conference on Science and Innovation Policy, and the 2013 Conference of the American Evaluation Association in Washington, DC were especially helpful.

Important data to support the quantitative analysis of real options and portfolio effects were provided by Robert Fri of Resources for the Future, Dickson Ozokwelu of the U.S. Department of Energy, Karen Jenni of Insight Decisions, LLC, and Robert Lanza of ICF International. Research assistance was provided by Philip Gilman and Nathan Cogswell.

The excerpt from "Characterizing the Relationships among a Set of Research and Developed Projects Sponsored by the U.S. Department of Energy" that appears in Chapter 4 is reproduced with the permission of ICF Resources, LLC. The image that appears as Figure 4.1 is reprinted with the permission of the National Academies Press.

Lastly, I am indebted to my family—Teri and Megan—for their continued support during this project.

DOI: 10.1057/9781137542090.0004

1

Research and Development: Opportunities and Challenges

Abstract: *Chapter 1 describes the tension between the goals of policymakers who seek to make research and development (R&D) a cornerstone of national economic progress and the reality that R&D programs operate in a complex environment where program design and implementation is challenging and success is difficult to measure. The chapter also summarizes how the rest of the book offers insights and tools to improve the opportunities for success in government R&D programs.*

Keywords: DOE energy efficiency programs; Government research and development; R&D management; R&D portfolio valuation; R&D risks

Linquiti, Peter D. *The Public Sector R&D Enterprise: A New Approach to Portfolio Valuation.* New York: Palgrave Macmillan, 2015. DOI: 10.1057/9781137542090.0005.

The global recession that began in 2008 had a profound effect on investments in research and development (R&D). The yearly growth rate in R&D spending by both the public and private sectors in developed countries fell to half of pre-recession levels (OECD, 2014a). Private R&D spending has since shown signs of recovery, but public spending has lagged behind. That notwithstanding, most governments around the world—not just OECD members—continue to affirm a goal of, at a minimum, maintaining R&D spending, or for most countries, of increasing R&D spending significantly (OECD, 2014b). China, for example, is on track to become the world's largest R&D performer by 2020 if current trends persist (OECD, 2014a).

Such governments see R&D, and innovation policy more broadly, as a prerequisite for dealing with daunting societal challenges such as slow economic and employment growth, the healthcare needs of ageing residents, the transition to a low-carbon economy, and in some countries, rapid population growth. At the same time, R&D is also seen as a means of exploiting opportunities in fields like information and communication technology, biotechnology, nanotechnology, advanced manufacturing, "big data," agriculture, and national defense. For these reasons, countries tend to view fostering R&D as an integral and vital component of public policy (Link & Vonortas, 2013).

The United States is no exception. In 2011, a total of about $429 billion was spent by the public and private sectors on R&D (National Science Foundation, 2014b) and President Obama has called for even more R&D spending (Executive Office of the President, February 2011). When it comes to government spending, the public sector R&D enterprise is a large one; the Federal government spends $130 billion per year on R&D.[1]

Despite the size of this investment, however, the Science of Science Policy Task Group concluded that "the rationale for specific scientific investment decisions lacks a strong theoretical and empirical basis" (ITG, 2008, p. 1) and the White House has told agencies to "improve management of their R&D portfolios and better assess the impacts of their science, technology, and innovation investments" (Executive Office of the President, 2010, p. 2). Such concerns reflect a global trend of increasing demands from elected officials that public R&D administrators demonstrate the return on their investment of taxpayer funds (OECD, 2014b).

Responding to such demands, however, is often difficult because public sector R&D projects are not typically selected or managed with

DOI: 10.1057/9781137542090.0005

quantitative methods. Instead, qualitative review is the norm because policymakers often believe that available tools undervalue the benefits of R&D investments, especially those where potential payoffs may be far in the future (Vonortas & Desai, 2007). Several investigators have offered more robust quantitative techniques to supplement qualitative judgments during R&D portfolio formation, valuation, and management (Vonortas & Desai, 2007; Tassey, 2003; Casault, Groen, & Linton, 2013).

This book continues along the same lines. My goal here is to make a methodological contribution focused primarily on the process of forming and valuing portfolios of R&D projects.[2] In a study of 39 R&D initiatives, the National Research Council concluded that "how programs were organized and managed made a real difference to the benefits that were produced" (NRC, 2001, p. 6). My hope is that this exploration of the topic will, at a minimum, clarify our understanding of R&D programs, and at best, increase the likelihood of their success.

R&D programs exist in a complex environment, and successful programs must be thoughtfully designed and well executed. Before diving into the topic of portfolio valuation, therefore, I start in Chapter 2 with a broad overview of the public sector R&D enterprise, particularly as it operates in the United States. I assume the reader has had little exposure to these topics and have structured Chapter 2 as a primer-like review of the basic structure and operation of Federal R&D programs.

The discussion in Chapter 3 focuses on the process of deciding which R&D projects should be funded and how they can be combined into a coherent portfolio. I look at management practices typically used in the private sector, and then summarize similar efforts in government R&D programs. With this foundation in place, I turn to original research in Chapter 4 and present the methodology and results of an analysis of a portfolio of projects sponsored by the Department of Energy between 2002 and 2004.

As explained in Chapter 5, my analysis of these projects suggests that traditional methods of valuing R&D investments significantly understate their benefits. In addition, I demonstrate that the typical government practice of evaluating individual R&D projects in isolation, without regard to potential relationships among them, makes it almost impossible to characterize the risks being taken with taxpayers' money. Finally, I conclude that proactive management of government-sponsored R&D, including interim performance reviews that can lead to project

DOI: 10.1057/9781137542090.0005

expansion, substantive redirection, or early termination, can improve outcomes as compared to a hands-off style of management.

Finally, while Chapter 4 demonstrates the feasibility of more rigorously quantifying the benefits of R&D projects, and the risks of the portfolios they comprise, I observe in Chapter 6 a few areas for improvement and briefly sketch methods for moving forward.

Notes

1 This spending can be broken into categories. Basic research is the study of scientific phenomena without direct regard for the practical application of new knowledge that may be developed. Conversely, applied research attempts to generate new knowledge in order to meet a specific need. Development entails efforts to produce devices, systems, and methods to put new knowledge into practical use (AAAS, 2013).

2 Some of the material in this book, especially in Chapters 3 and 4, draws directly from my dissertation (Linquiti, 2012b).

DOI: 10.1057/9781137542090.0005

2
A Primer on the Public Sector R&D Enterprise

Abstract: *The US government spends $30 billion per year on applied, nondefense R&D. Chapter 2 provides a holistic overview of how this R&D enterprise operates. Context is set with a review of the political, economic, and institutional forces that shape R&D programs. A logic model framework is used to describe the inputs, activities, outputs, outcomes, and impacts of public sector R&D. This expansive approach makes the subject more complex, but protects the reader from analytic myopia—a failure to recognize all of the drivers of success and failure. The primer aspires to provide readers who have a limited background in the field a readily accessible entry point to the world of government R&D programs.*

Keywords: economics of R&D; government research and development; logic models; politics of R&D; R&D program evaluation; technology diffusion

Linquiti, Peter D. *The Public Sector R&D Enterprise: A New Approach to Portfolio Valuation*. New York: Palgrave Macmillan, 2015. DOI: 10.1057/9781137542090.0006.

This chapter describes a framework for thinking holistically about publicly supported R&D investments that are intended to stimulate the development and deployment of new technologies. My audience is students, public administrators, and others who seek a basic primer on the design, implementation, and evaluation of these types of public sector R&D programs. The focus here is on the United States and, in particular, on Federal (as opposed to state, regional, or local) programs.

Before beginning, I should point out that the framework does not address basic research or defense R&D. Analysis of basic research initiatives, intended to generate new knowledge without regard to potential commercial application, brings with it additional theoretical, practical, and policy considerations that go beyond those typically relevant to applied R&D. Rather than extend the proposed framework to address basic research, I instead focus more narrowly (and I hope more usefully) on applied R&D.

As for defense R&D, the presence of a single, large customer (e.g., a national defense ministry) means that the market dynamics for new defense technologies are very different than for nondefense technologies that compete in a market with multiple private buyers and sellers. Defense R&D also occurs in an environment marked by a much closer relationship between producer and customer than is the case in commercial markets where arms-length relationships are far more common. Finally, attaching a monetary value to the benefit of enhanced national security—presumably an outcome of successful defense R&D—would be a daunting analytic challenge.

Even after excluding basic research and defense R&D, considerable sums of money are still at stake. The Federal government spends about $30 billion on nondefense, applied R&D in fields such as health, energy, agriculture, transportation, and the environment (AAAS, 2013).

2.1   Overview

The framework that I describe here aims to identify a comprehensive set of policy, political, technical, and market factors that may affect the success or failure of an R&D initiative. The framework is multidisciplinary, meaning that it includes economic considerations such as market failures and the macroeconomy, political considerations such as earmarks and lobbying, and institutional behavior such as interagency coordination and bureaucratic rules. Why take an expansive approach? Doing so makes the

DOI: 10.1057/9781137542090.0006

subject more complex, but a core tenant of the modern discipline of public policy analysis is that no single mode of inquiry offers a definitive vantage point from which to examine a public challenge; instead, complementary insights from multiple perspectives yield a more robust understanding (deLeon & Martell, 2008; Stone, 2012). What's more, a narrow perspective, perhaps focused on only one element of the R&D enterprise, runs the risk of analytic myopia—a failure to recognize all the important factors that ultimately determine whether a new program succeeds or fails.

The framework is not an explanatory model per se; instead, it is conceptual, intended to offer administrators a way to locate their work in a broader context and identify key linkages among disparate factors. Not all of the factors identified in the framework will apply to all R&D initiatives. Rather, the framework can be used as a checklist to ensure that important factors are not overlooked during the design, implementation, and evaluation of public R&D programs. Of course, depending on the specific circumstances of a particular program, other factors may also be relevant.

The framework is structured as a logic model, so named because of the implicit chain of if-then statements embedded in it. Logic models are a common tool in the field of program evaluation and provide a systematic way of characterizing public (and nonprofit) initiatives (Newcomer, Hatry, & Wholey, 2010). Logic models draw causal linkages from the initial inputs for a program, through its operations, and on to its final impacts. Mediating factors that affect the operation of the causal relationships can also be depicted in a logic model. Because of its comprehensive nature, a logic model provides a useful way of describing public sector R&D initiatives. For an excellent summary of logic models applied to the R&D process, see Jordan (2013).

My proposed framework for analyzing public sector initiatives aimed at applied R&D is presented in Figure 2.1 The five core components of the model are arrayed from left to right, beginning with inputs used in the R&D program and ending with the ultimate impacts of the program on society writ large. The framework also identifies three sets of mediating factors which, though they exist outside the R&D program itself, may have a profound effect on it. Finally, the framework notes the potential for program evaluation, a process by which systematic efforts are made to understand the results of the program. The remainder of this chapter offers some observations on each of the five components of the logic model, as well as on the three sets of mediating factors and on the program evaluation process.

DOI: 10.1057/9781137542090.0006

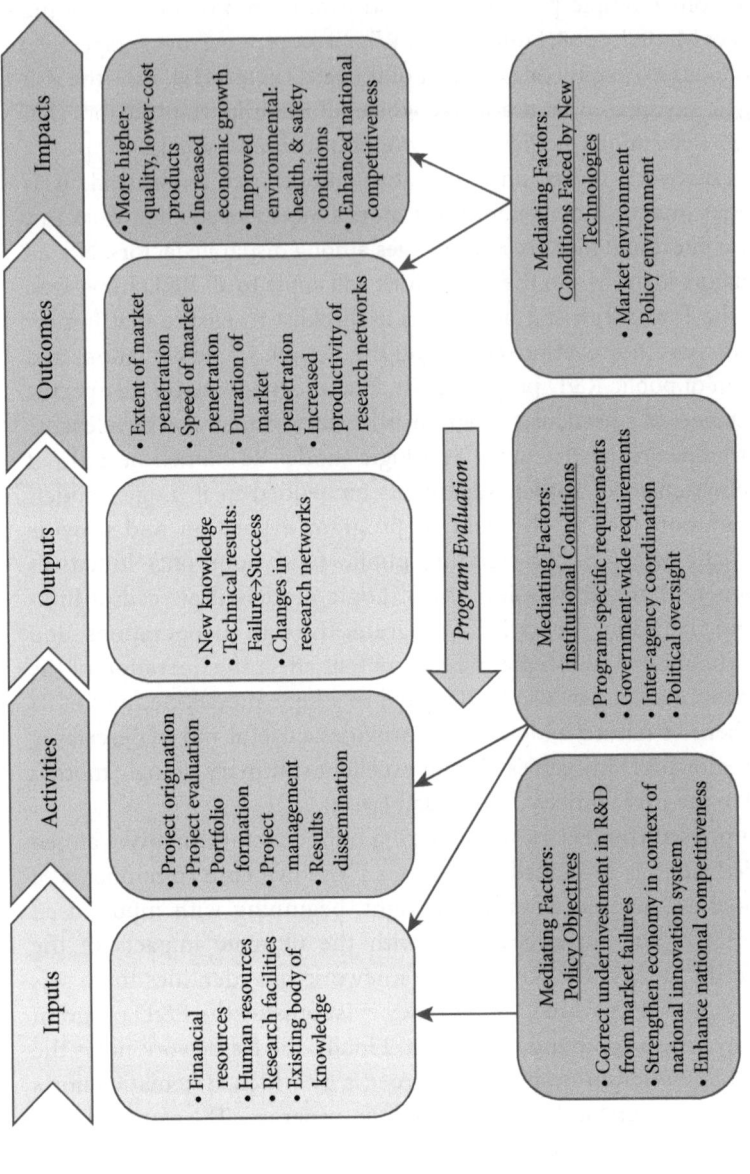

FIGURE 2.1   *The public sector enterprise for applied research & development*
Source: Adapted from Linquiti (2012b).

2.2   Mediating factors: policy objectives

Before describing the core components of the typical publicly funded R&D initiative, it is helpful to review the reasons why policymakers might create such an enterprise in the first place. Rarely is a single rationale put forward; rather, multiple arguments are usually marshalled to make the case for a new R&D program. The 2009 testimony of Secretary of Energy Steven Chu before the Senate Committee on Energy and Natural Resources is a case in point:

> I am here to speak for clean energy R&D. Investment in energy R&D will drive innovation across the economy and maintain American competitiveness. It will create jobs and entire new industries. And it is vital for meeting the energy and climate challenge.... It is imperative that government provide R&D funding, especially at the front end when private investments would not recoup the full value of the shared social good or when a new technology would displace an embedded way of doing business.... Federal R&D also builds the human, physical, and technological capital needed to perform breakthrough research. (pp. 1–2)

Understanding the rationale(s) behind an R&D program often yields considerable insight into its structure and operation. Accordingly, this section lays out three arguments often advanced in support of publicly funded R&D. A quick caveat: I don't aim to evaluate the validity of each rationale (doing so would make this chapter unmanageably long), but simply to describe it and provide one or two examples.

2.2.1   Correct market failures

One rationale for public sector R&D programs is that the private sector cannot be expected to routinely invest sufficient resources in R&D, thereby causing markets to "fail." By way of summary, the typical argument is that freely competitive markets will generate a suboptimally low level of R&D activity (Nelson, 1959; Arrow, 1962; Griliches, 1992). Because of the intangible nature of technical and market knowledge and inherent limitations in the legal mechanisms used to protect intellectual property, it is typically very difficult for the developer of such knowledge (i.e., the funder of the R&D) to fully appropriate all the benefits of its investment, as some of those benefits "spill over" to customers and competitors without compensation to the original R&D funder. If the funder *could* appropriate more of the monetary value of these benefits,

DOI: 10.1057/9781137542090.0006

then presumably it would be willing to invest more in R&D. The resulting systematic under-investment in R&D creates a rationale for government intervention to supplement private R&D spending.

Some scholars have suggested that this argument about insufficient private investment in R&D is most compelling in the case of basic research far removed from commercial markets and that it is less relevant for applied R&D where firms are focused on goods and services that can compete successfully in the market (Bernanke, 2011). Others, however, have argued that government-supported applied R&D in fields like aviation, nuclear power, computing, and satellites, has been instrumental in creating innovative technologies that would not otherwise exist (Ruttan, 2006). In addition, a review of other developed countries reveals a higher level of government support for, and coordination of, applied R&D, than is the norm in the United States (Atkinson & Ezell, 2012).

As Tassey notes, in addition to the "appropriability" problem, there are other reasons why firms may under-invest in R&D (2005). A project may be so expensive that a single firm cannot secure the resources to undertake it. Alternatively, a project—even one more likely than not to yield a positive return—may be both sufficiently risky and costly that should it fail, it could bankrupt even a large firm. Such an existential threat may dissuade all but the most aggressive firms from undertaking the project. In these cases, government intervention (and the investment of resources from millions of taxpayers) may create sufficient cost and risk sharing that the R&D project could be justified.

Other R&D projects may support public goods (e.g., testing protocols and measurement methods for technologies upon which many firms depend). The technical risk may not be high, but no firm wants to invest in the development of the standard, knowing that it will be unable to recoup its costs and that it lacks the credibility to establish a standard upon which its competitors would be willing to rely. The National Institute of Standard and Technology funds projects in this arena and a recent review of 20 such projects reported that benefits routinely exceed costs, typically by a wide margin (Link & Scott, 2012).

A final example of a market failure that may be corrected by public R&D comes from the externalities generated by existing technologies, especially those that create environmental, health, or safety concerns. Economists define an externality as an impact on one party created by another party, the value of which is not fully reflected in the price of an associated commercial transaction. For example, air emissions from a factory may cause adverse health effects in people who are not compensated for their injury. If the producers and users of technologies do not incur the full social cost of their activities, their incentives to develop and deploy less harmful technologies will be curtailed (Bozeman & Rogers, 2001). For example, emissions of greenhouse gases are not typically regulated or taxed in the United States. Accordingly, without a "price on carbon," the rewards for technical innovations that reduce emissions are lower than they otherwise would be (Jaffe, Newell, & Stavins, 2004).

Having identified a series of market failures that may lead to a suboptimal level of R&D investment, one might be tempted to conclude that policymakers simply need to spend more public funds on R&D to raise the level of R&D investment to an appropriate level. While this might be true, policymakers need to be mindful of the potential for "government failure," meaning that good intentions aside, government R&D managers can be challenged by the difficulties of deciding what types of R&D initiatives to support, how support should be delivered, and how to avoid simply displacing private funds that would otherwise have been invested in the same R&D activities (Feller, 2011). Ideological debates may also emerge about whether the government, as it supports some R&D efforts and not others, is attempting to "pick winners" from among nascent competing technologies, a task that some argue is best left to the private marketplace. For a good review of this debate, see Atkinson and Ezell (2012).

2.2.2   Strengthen the economy

Arguments in support of public sector R&D programs often cite a need to enhance the overall national economy, either as part of an ongoing to effort to create and maintain economic prosperity, or as a one-off measure to stimulate the economy in times of economic contraction. I discuss each of these two rationales in turn below.

Beginning in the 1950s, as macroeconomists studied the growth of developed economies, they detected a share of national economic growth that could not be readily explained by increases in the quantity or quality of factor inputs (like capital and labor) and ultimately identified technical change and productivity growth as key drivers of national economic performance (Abramovitz, 1956; Solow, 1957). In turn, these findings have spurred almost six decades of analysis of the returns to R&D—including both the private returns to R&D investors and the social returns, inclusive of beneficial spillovers, to society writ large

(Feller, 2011; Tassey, 1997). A strand of this research has also specifically investigated the returns to government-funded R&D and its effect on private R&D. While substantial methodological and empirical uncertainties remain, two extensive literature surveys have recently suggested that there are usually significant returns to R&D, that social returns are routinely larger than private returns, and that while government-funded R&D may sometimes simply displace private R&D funding, it often does not (Hall, Mairesse, & Mohnen, 2009; Zuniga-Vicente, Alonso-Borrego, Forcadell, & Galan, 2014).

As mentioned in Chapter 1, several countries have pursued increased R&D investment as an explicit national growth strategy (Link & Vonortas, 2013; Goel, Payne, & Ram, 2008). For example, both the European Union and the United States have articulated a goal of combined private and public investment in R&D equal to at least three percent of Gross Domestic Product (GDP) (European Commission, March 19, 2014; Executive Office of the President, February 2011). The United States is close to the target, with 2.85 percent of GDP going to R&D, while the European Union invests 1.94 percent of GDP in R&D (National Science Foundation, 2014b, pp. 4–19). In both the United States and the European Union, about a third of R&D funding comes from government (National Science Foundation, 2014b, pp. 4–21; Eurostat, 2014).

One brief caveat is important here. Scholars working in this field have noted that government R&D investments intended to stimulate national economic growth are always made within the context of a larger enterprise typically referred to as a national innovation system, or NIS. Variously credited to Freeman, Nelson, and Rosenberg, and Lundvall, the NIS paradigm captures the complexity of interactions that drive innovative behavior—the cornerstone of the relationship between R&D and economic growth (Edquist, 2005). Analysts typically identify four components in an NIS: (1) the scientific, technical, and business process knowledge upon which it is based; (2) the actors within the system, including private firms, the government, universities, and nonprofit institutions; (3) the relationships among these actors, including both market competition and mutually agreed collaboration; and (4) the "rules of the game," including legal requirements, intellectual property protection, tax policy, industry standards, cultural norms, and so on, that shape interactions among the actors. What's more, Malerba notes that these four components are rarely if ever static, but instead are in a state of constant flux, often co-evolving over time (2002).

DOI: 10.1057/9781137542090.0006

Understanding R&D not in isolation, but as a larger system of innovation, has become common in the academic and policy literature (OECD, 2014b; Flanagan, Uyarra, & Laranja, 2011; Aghion, David, & Foray, 2009; Arnold, 2004; Smith, 2000). While there are important differences of approach within this literature, taking a broad, systems-oriented view has become the norm. In turn, there is little doubt that variations in the character and operation of a country's national innovation system can have a significant effect on the results of government R&D investments. In short, R&D is indeed an important element in economic growth, but it is not the only one.

Closely related to policymakers' desire to enhance national economic growth is a desire to improve the long-run employment situation, in terms of both the number of jobs and the wages paid to workers. These aspirations are evident in Secretary Chu's remarks noted earlier, as well as in President Obama's Innovation Strategy, the second sentence of which argues that innovation can "create the jobs and industries of the future" (Executive Office of the President, February 2011, p. 1). To be sure, government support of R&D is only one among many policy measures in the President's strategy, and employment growth is only one among many anticipated outcomes of the strategy. That notwithstanding, claims about the prospect of more and better jobs are pervasive in today's policy debates about government R&D investments. Whether such claims are meritorious, however, is not clear. As Link and Scott point out, scholarly investigation of the direct link between "public R&D subsidies and the employment effects of the subsidized research is still nascent" (2013).

Independent of the desire to use ongoing R&D investments to expand the economy, and increase wages and employment, there are also times when policymakers are especially concerned about conditions like high unemployment or declines in GDP. During a recession, policymakers may endeavor to manage taxes and spending in ways that stimulate the economy.[1] Doing so usually does not include significant investments in R&D, but the American Recovery and Reinvestment Act (ARRA), also known as the Stimulus Bill, is an exception. Enacted in February 2009 in the face of high unemployment and a rapidly contracting economy, ARRA contained about $830 billion in stimulus measures, including tax cuts, direct payments to individuals, and increased spending in government programs (Congressional Budget Office, 2014). Relevant for our purposes here is that ARRA provided $19.2 billion in one-time funding

DOI: 10.1057/9781137542090.0006

for Federal R&D programs (National Science Foundation, 2013), of which approximately $7.7 billion went to applied, nondefense R&D in Fiscal Years 2009 and 2010 (National Science Foundation, 2012).[2] In an effort to quickly move the stimulus funds into the economy, most agencies simply expanded existing R&D programs by calling for and funding additional grant proposals.

This was not, however, true in all cases. ARRA provided the required funds to launch the Advanced Research Projects Agency-Energy (ARPA-E), a new energy R&D program modeled on a successful US Department of Defense program. Created on paper by the 2007 America COMPETES Act, but left unfunded, ARPA-E received $400 million as initial funding under ARRA and began operations in 2009. Congress has subsequently provided ARPA-E with continuing funding as part of the ongoing appropriations process (Bonvillian & Van Atta, 2011).

2.2.3   Enhance national competitiveness

In a tradition that dates back at least as far as the 1957 launch by the Soviet Union of the Sputnik satellite, and claims made by then-Senator John Kennedy about being on the wrong side of a nuclear missile gap, policymakers have often argued that investments in R&D are needed to protect America's position in the world (Preble, 2003; National Academy of Science, 2007). The debate continued in the 1970s and 1980s as economic growth in the United States and many Western European countries slowed, and Japan demonstrated substantial economic and technological success. As Nelson and Rosenberg put it, there emerged a "new spirit of…'technonationalism'" (1993). Such sentiments continue to inform US policies, as evidenced by the title (and content) of the American COMPETES Act—passed in 2007 and reauthorized in 2010. Similarly, Energy Secretary Chu's testimony quoted earlier cites "American competitiveness" as one reason to invest in energy R&D. A unifying theme here is that the United States, as a matter of policy, ought to go to great lengths to ensure that it does not come out a loser in the competition.

Most debates about "national competitiveness" do not, however, consistently define the term in a precise and unambiguous fashion (Carayannis & Grigoroudis, 2014). Instead, multiple policy problems *and* multiple policy solutions are often brought under the umbrella concept of competitiveness. A 2007 study, "Rising Above the Gathering

DOI: 10.1057/9781137542090.0006

Storm" (RAGS), offers a case in point. After expressing concerns about disconcerting trends in R&D investment, corporate innovation, K-12 science and math education, high-tech employment, and advanced manufacturing, the National Academy of Sciences explicitly posed the question "What if the United States is not competitive?" (p. 205) and then answered by noting that while the United States has:

> led the world [in science and technology] for decades, the world is changing rapidly and our advantages are no longer unique. Without a renewed effort to bolster the foundations of our competitiveness, it is possible that we could lose our privileged position over the coming decades. For the first time in generations, our children could face poorer prospects for jobs, healthcare, security, and overall standard of living than have their parents and grandparents. (p. 223)

The National Academy study offers a number of policy proposals, one of which is a substantial increase in R&D funding.

Another aspect of national competitiveness focuses on US national energy security. Disruptions of Middle East oil supplies in both 1973 and 1979 had significant impacts on the US energy system, causing shortages of gasoline, turmoil in energy-dependent industries, and extreme price volatility, and began a long tradition of Federal efforts to reduce reliance on imported sources of energy, along with more extensive deployment of military forces in energy-rich regions of the world. A desire for national energy independence has explicitly motivated a number of Federal R&D programs (NRC, 2001).

If our aim is to carefully delineate the rationales that motivate a particular R&D initiative, we might be tempted to dismiss the concept of competitiveness as too amorphous as to be useful. Instead, we might focus on the underlying concerns of trade balances, job loss, technological innovation, market failures, energy price volatility, and so forth. Doing so would not be wrong, but it would miss something I believe is important. The fact that competitiveness is so often invoked in defense of science and technology policies suggests that the term has a symbolic meaning that goes beyond the tangible benefits of a specific set of policies. The term seems to resonate with multiple audiences, including politicians, the public, industry, and the media. After all, if we don't win the competition, then we lose it. And which government agency or politician wants to be seen as a loser, especially of what the RAGS study calls a "privileged position?"

DOI: 10.1057/9781137542090.0006

2.3   Model component: inputs

Having reviewed the potential rationales that may create the context for, and motivate the operations of a public R&D program, I now turn to the first of the core components of the logic model, the inputs into the process. In brief, there are four essential inputs in any R&D initiative, three tangible, and one intangible. Although perhaps obvious, these inputs are worth a brief review.

First, sufficient financial resources must be available to cover the cost of, among other things, personnel (such as salaries, benefits, and overhead expenses), facilities (such as offices, labs, warehouses, and testing areas), equipment (such as fabrication equipment, testing instruments, reference materials, and computing technologies), and raw materials (that may be consumed during experimentation or needed to build or operate bench- or pilot-scale prototypes). Not only is the aggregate amount of funding important, so too is the timing of its availability. R&D projects are often executed over a multiyear period, with costs often incurred irregularly during the project. If sufficient funding is not available at the right time, the substantive work of the project may be disrupted. In addition, funding may come from multiple sources, especially in government-sponsored R&D programs. Cost-sharing between public and private sources is the norm in applied government R&D programs in the United States. What's more, multiple government agencies and private parties may pool funding to support a single project or program.

Second, human resources are a critical input to any R&D initiative. Of course, the key investigators and the technical support staff must have the right mix of scientific and engineering expertise to successfully execute the project. And, at appropriate points in the project, staff with an understanding of business conditions, customer needs, and cost engineering, need to be involved. Also important are the managerial skills of key personnel who are responsible for breaking the research project down into its component tasks, scheduling those tasks, and ensuring that resources are available at the right time while taking care that specific resources (either personnel or facilities) are not overburdened in any given time period. Project managers in government R&D programs must also possess the training and experience needed to comply with unique public sector regulations related to procurement, contracting, personnel, and budgeting. In addition, turnover of key personal can also affect the

DOI: 10.1057/9781137542090.0006

prospects for the success of a project; the tacit knowledge gained during the early phases of a project can create a steep learning curve for a new team member who joins a project mid-stream.

Finally, when a single research team comprises staff who work for different organizations (e.g., two universities or a private firm and a government lab), additional complications may arise (Bozeman, 2000). Such complications could include tensions over reporting relationships (does the staff member answer to the principal investigator, even if he/she works for a different employer?), treatment of confidential information (how does a university researcher accustomed to academic freedom deal with a commercial partner, anxious to protect what may become trade secrets?), and potential changes in project scope (faced with project termination due to unexpectedly poor research results, how does the untenured academic researcher react as compared to his/her private sector colleague who will be quickly re-assigned to the firm's next research priority?).

Third, R&D programs require suitable facilities in which the R&D can be performed. In the United States, a large majority of the $31.4 billion in applied research paid for by the Federal government is not actually performed in facilities fully owned and operated by the government. Instead, as shown in Figure 2.2, non-Federal institutions perform 74 percent of the federally funded applied R&D (National Science Foundation, December 2013).

In some cases, the funding vehicle is a Federally Funded R&D Center (FFRDC) sponsored by a Federal agency and operated in partnership with a non-Federal institution. About 17 percent of Federal applied research funds flow through FFRDCs, which are either owned and operated by the non-Federal partner, or owned by the Federal government but operated by the partner. FFRDCs typically involve close relationships between the government and the non-Federal partner, with more sharing of facilities, personnel, data, and intellectual property than is the norm in the typical government–contractor relationship. As of May 2014, there were 40 FFRDCs in operation; examples include the Los Alamos National Laboratory, the Center for Naval Analyses, the Science and Technology Policy Institute, the Judiciary Engineering and Modernization Center, and the Jet Propulsion Laboratory (National Science Foundation, 2014a).

While the academic sector tends to focus on basic research and the industrial sector tends to focus on applied R&D, this is far from a

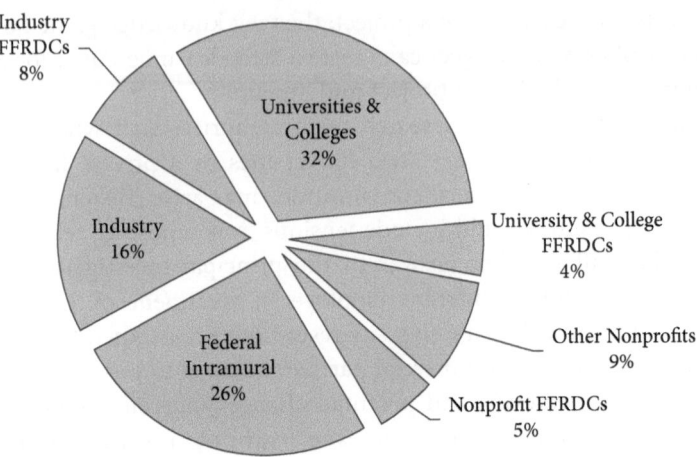

FIGURE 2.2   *Entities performing federally funded applied research in 2012*[a]

Source: National Science Foundation (December 2013).

Note: [a] Excludes funding for basic research and for development, but includes applied research in the defense sector.

universal pattern, and in recent years, academic participation in applied research has been increasing (Goel & Rich, 2005). For example, universities and colleges—both as independent entities and as participants in FFRDCs—are the single largest performer of Federal applied research (36%) and, in turn, Federal monies comprise about 61 percent of all university and college R&D spending (National Science Foundation, 2014b, p. 5–10). Industry performs 24 percent of Federal applied research work (either on a stand-alone basis or as part of an FFRDC). Similarly, nonprofit institutions are responsible for the remaining 14 percent.

As shown in Figure 2.2, only about 26 percent of Federal applied research monies is spent at facilities within the government itself. Examples of public organizations with such facilities (not necessarily limited to applied research) include the Agricultural Research Service, the US Geological Survey, the Environmental Protection Agency's Office of Research and Development, the National Institutes of Health, and several of the Department of Transportation's modal agencies (Sargent, 2014).

The fourth input to the R&D process is intangible, but no less important. All R&D initiatives take as a point of departure the existing base of knowledge in the relevant fields. Before they are made, technical and

DOI: 10.1057/9781137542090.0006

scientific breakthroughs reside in what one author has, in the context of innovation, called the "adjacent possible" (Johnson, 2011).[3] If the objective of an R&D initiative is simply too far afield of the existing state of knowledge (i.e., it is not adjacent to today's understanding of the phenomenon of interest), success may prove impossible. Conversely, whether researchers can break through to the adjacent possible depends, in part, on the degree to which they are able to readily access the full body of existing relevant knowledge. If important theory, methodologies, data, or findings exist somewhere in the world, but are not available because of intellectual property protections, because they are not internet-accessible, or because they are written in a language not easily read by the research team, then this knowledge cannot inform the R&D process. A key input, therefore, to all R&D initiatives is an accessible body of knowledge that represents the state of the art in all relevant fields, upon which the research project can then build.

2.4   Model component: activities

Given these four basic inputs to an R&D program—money, people, facilities, and knowledge—the next defining characteristic of a program is how those inputs are used to facilitate a set of activities within the program itself. Such activities, which constitute the second element of the R&D logic model, focus on the conduct of the research over its full lifecycle, from start to finish. Although intertwined, for ease of exposition, it is helpful to think of these activities as comprising five sequential steps.

The first step is project origination, the mechanism by which proposed R&D projects are initially created and identified as potential candidates for public funding. The norm, at least at the Federal level, is an open, competitive process in which the availability of R&D funding is announced and interested parties are invited to submit project proposals. Some Federal agencies first ask for concept papers, and then encourage full proposal submissions only from applicants whose concept papers are the strongest. The scope of funding announcements can vary widely. ARPA-E, for example, issues both broad solicitations for "any idea that has the potential to produce game-changing breakthroughs in energy technology" and narrowly focused solicitations, such as a call for proposals to "develop technologies that utilize abundant domestic natural gas as fuels for passenger vehicles" (2013, p. 2).

DOI: 10.1057/9781137542090.0006

In some cases, prior to issuing a solicitation, the funding agency identifies a suite of complementary technologies that must be developed in order to achieve a particular breakthrough. This process is sometimes referred to as technology road-mapping. Funding announcements may then seek proposals for specific components of the technology roadmap. In other cases, considerable discretion is left to individual proposers to decide the technical area to be addressed and the methods to be applied.

Solicitations also describe the legal arrangements—typically a grant, cooperative agreement, or contract—that will guide the research project, as well as the types of entities that may submit a proposal. Some solicitations are open to all potential respondents while many, especially those relying on grant mechanisms, are limited to nonprofit entities like universities and research institutes. When for-profit firms are eligible to compete, there may be size limitations (e.g., only small business may participate) or teaming requirements (e.g., only firms partnered with a nonprofit entity may participate).

The second major activity in an R&D program is project evaluation, in which the merits of individual R&D project proposals are assessed. Proposal evaluation tends to be a formal process in most agencies, especially because considerable sums of taxpayer money may be at stake. The funding solicitation usually spells out an explicit set of criteria to be used to assess proposals. Technical merit is always assessed, and sometimes costs are also considered. Proposals are typically evaluated by a group of reviewers, who may be government employees or may be drawn from the private sector or academia. Reviewers identify the strengths and weaknesses of each proposal, and then usually confer to agree on ratings for the most promising proposals.[4] On the basis of this review, the agency may go back to some or all of the researchers to pose questions or suggest revisions. Final proposals are then submitted, and subjected to additional review.

The third key activity is portfolio formation, during which program managers select a set of projects for funding. In some cases, the selected portfolio is little more than a collection of the top-ranked proposals, where each proposal has been considered only as stand-alone activity. While this approach ensures that the best proposals are selected, it may result in a set of projects that aim to explore the same phenomena with similar methodologies, while leaving other topics of interest unaddressed or promising methodologies unapplied, simply because they were not included in any of the top-ranked proposals. In other cases,

DOI: 10.1057/9781137542090.0006

program managers try to select a coherent collection of interrelated projects, intending to ensure that complementary research questions are posed, diverse data sources are developed, and multiple methodologies are tested. While doing so may increase the chances that the research results—as a whole across all projects—will form a larger and more coherent body of new knowledge, it may also entail passing over high-quality proposals in favor of lower-ranked proposals. Section 3.3 delves into the process of portfolio formation in considerably more detail.

The fourth key activity is project execution and management during which the research is conducted by the proposal team. The role of the government in the day-to-day project execution depends on the legal arrangements specified in the solicitation. A grant award is the most hands-off arrangement; under the typical grant, government/researcher interactions may be limited to annual reports (or to an annual meeting of grantees) and a final report. This approach reflects the relevant statutory requirement that, under a grant, there should be no "substantial involvement" between the agency and the grantee (31 USC 6304).

A cooperative agreement, on the other hand, is used when "substantial involvement" between the government and the researcher *is* expected (31 USC 6305). Cooperative agreements are used in many applied R&D programs to allow more government influence over the nature and direction of the funded research work and to facilitate a greater flow of information about interim research results to government program managers, and in some case, to other research teams in the program.

Finally, the government may procure research services under a contract which directly specifies the research questions to be addressed, how the work is to be done, the nature of the deliverables to be produced, and the schedule for the work. Under most contracts, the government retains considerable discretion to modify the scope of work, reset priorities, or re-allocate funds among tasks. While a contract gives significant control to the government to focus the research on its highest priorities, it also requires the government agency to have the technical expertise to initially frame the research project and then to manage it closely.

In addition, government contracts for R&D can take different forms. Under a cost-plus-fee contract, the private research entity is reimbursed for all of its costs and without regard to the outcome of the research. In contrast, under a fixed price contract, the payment is not formally linked to the contractor's cost, but instead reflects the price it bid during the contract award process. Under a fixed price contract, if the cost of

DOI: 10.1057/9781137542090.0006

executing the R&D project is less than expected, the contractor retains the difference as profit but conversely if costs are higher than planned, the contractor must use its own funds to complete the project. Depending on the nature of the contract, the payment under a fixed price contract may or may not depend on the extent of technical success. Finally, a variety of incentive-based contracts exist under which the risks and potential profits of technical success are shared between the government and the contractor. The literature suggests that no one contract mechanism is optimal in all situations; instead, the best approach depends on the magnitude of the R&D cost, the intrinsic uncertainty of the results, the potential benefits of a successful research result, and the number of likely bidders (Goel, 1999).

Under cooperative agreements and contracts, government managers typically can subject ongoing projects to reviews (sometimes called stage-gate reviews). During such reviews, decisions may be made about whether to continue the research, to expand it or scale it back, to re-orient it, or to terminate it. With a grant-funded R&D project, however, the government has almost no authority to redirect or terminate the work based on interim research results or changing market conditions.

As explained in Chapter 3, stage-gate reviews are a common practice in private sector R&D programs, but have traditionally been less prevalent in government programs. An exception to the pattern is ARPA-E, which touts its "hands-on engagement with awardees" and its ability to redirect projects that are not performing as hoped (ARPA-E, 2013, p. 3).[5] As of February 2014, ARPA-E had canceled 18 of 362 funded projects prior to completion, either because of technical problems or because the prospects for commercialization of a new technology had changed (St. John, 2014).

The fifth major program activity entails efforts by either the researchers or the program administrator to disseminate the research results to wider audiences. Governments typically make doing so a high priority for publicly funded R&D projects (OECD, 2014b). The nature of such efforts usually depends on the result of the R&D initiative. If the result is an innovation that is close to being ready for the marketplace, the focus is on enhancing the chance of commercialization. This process is often referred to as technology transfer, although several meanings are ascribed to this phrase within a "voluminous, multidisciplinary literature" (Bozeman, 2000, p. 627). Government-sponsored technology transfer can take many forms, ranging from involving industry (either in a consortium or as individual firms)

DOI: 10.1057/9781137542090.0006

in the R&D itself to licensing new technologies to private firms to working with end-users to stimulate demand for the new technology (Brown, Berry, & Goel, 1991).

ARPA-E goes further and offers consulting services to its research teams on how to commercialize the results of their research and "facilitates relationships with investors, government agencies, small and large companies... to move awardees to the next stage of their project development" (ARPA-E, 2013, p. 4). Several frameworks have been developed for evaluating the effectiveness of technology transfer programs (Georghiou & Roessner, 2000; Kingsley, Bozeman, & Coker, 1996; Brown, Berry, & Goel, 1991).

In the case of an R&D initiative that yields valuable technical progress, but not a new technology reasonably close to commercialization, the public sector R&D manager faces a dilemma when it comes to information dissemination. The results of such R&D can certainly be compiled and published, thereby entering the collective body of knowledge, but evidence suggests that private firms place relatively little value on these formal, arms-length efforts to share technical knowledge, preferring instead to be involved in additional contract or collaborative research sponsored by government (Bozeman, 2000). Of course, follow-on R&D of this type entails additional cost and oversight responsibility for the government program.

2.5   Mediating factors: institutional conditions

Government oversight creates a unique environment for a public R&D program, which in turn affects the inputs available to it and the activities that it can undertake. Four characteristics of this environment are important: the authorities under which a specific program operates, the authorities generally applicable to all government programs, the relationship between any one R&D program and the activities of other government agencies, and the political realities typically faced by government R&D programs.

2.5.1   Program-specific statutory and regulatory requirements

The nature of a government program renders it subject to the legal authorities under which it is operated. In general, there are three types of authorities that are important for any individual R&D program:[6] the statutory authorization of the program itself, the appropriation of funds to operate the program, and the establishment of legally binding regulations that dictate some or all of its activities. To illustrate these three types of authority, it may be helpful to consider the example of the Advanced Technology Program (ATP), a Federal R&D program that operated between 1990 and 2007 and which was intended to help US businesses develop and commercialize new scientific discoveries and technologies.[7]

The agency in which an R&D program operates will almost certainly have been established by an act of Congress—often referred to as an agency's "organic act"—that lays out its structure, scope, and core mission. The ATP, for example, was initially created by the Omnibus Trade and Competitiveness Act of 1988, which renamed and restructured the National Bureau of Standards, created the National Institute of Standards and Technology (NIST), and established ATP as a program within NIST (Schooley, 2000). The relevant portion of the 1988 legislation included five pages of single-spaced text that prescribed, among other things, the types of projects and project sponsors that could be funded by ATP, the contracting vehicles to be used, the treatment of intellectual property, the role of an advisory committee, and reporting requirements for the program.

Over time, an agency's organic statute may be amended to add or eliminate responsibilities, restructure the organization, or adjust regulatory authorities possessed by the agency. In the case of ATP, the American Technology Preeminence Act of 1991 amended the original 1988 legislation to, among other things, direct ATP to focus on industry-led R&D efforts, to emphasize fields such as high-resolution information systems, advanced manufacturing, and advanced materials, and to no longer seek licensing fees from joint ventures (U.S. Congress, 1992). In 2007, Congress abolished ATP with the America COMPETES Act (U.S. Congress, 2007b).

A second way in which Congressional legislation directly affects the operations of a specific R&D program is through the appropriations process. Appropriations authorize the use of taxpayer funds from the Treasury to pay not only for Federal salaries and facilities, but also for the transfer of funds to non-Federal program participants. Legislative appropriations often include statutory language, or are accompanied by Congressional reports, that give further instructions on how the monies are to be spent. As Figure 2.3 shows, Congress initially provided ATP less than $50 million per year, but after a spike up to almost $350 million

DOI: 10.1057/9781137542090.0006

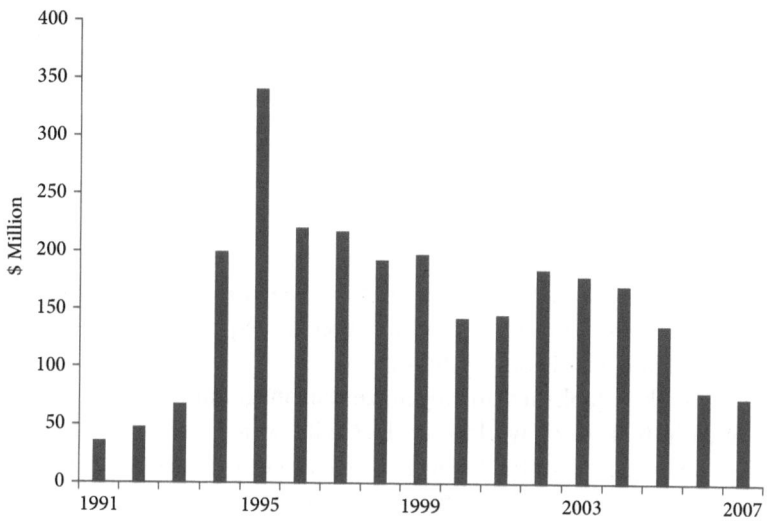

FIGURE 2.3   *Annual funding for Advanced Technology Program*
Sources: Schacht (2005, 2013).

in 1995, annual funding oscillated between about $221 million and $170 million for several years before falling to less than $80 million in the final two years of the program (Schacht, 2005; 2013).

The final legal authority that may affect the operations of a particular R&D program is regulatory in nature. Typically, Congress does not dictate all aspects of a program's operation and instead delegates to the administering agency the authority to further define the relevant requirements, which are then codified in the Code of Federal Regulations (CFR). This rulemaking process is subject to the Administrate Procedure Act, and typically includes publication of a proposed rule, consideration of public comments on the proposal, and promulgation of a final rule (Dudley & Brito, 2012). Regulations related to the operation of ATP were codified in Part 295 of Volume 15 of the CFR. They provide a level of detail about ATP activities not present in the underlying statutory language. The initial regulations governing ATP were promulgated in July 1990, and then subsequently revised three times (in 1994, 1997, and 1998).

In sum, as they endeavor to manage a Federal R&D program, administrators will find themselves constrained by the legal requirements of the organic statute, as amended, that authorized the program, by the annual appropriations process, including the amount of funding

DOI: 10.1057/9781137542090.0006

provided as well as Congressional instructions that accompany those funds, and by the regulations that have been previously promulgated by the program.

2.5.2   Government-wide statutory and regulatory requirements

Beyond these legal requirements for a given R&D program, there are two sets of requirements that apply to all Federal programs and which can be just as important as program-specific requirements: the Federal personnel and acquisition systems.

Originally established to minimize patronage and political interference with the Federal workforce, the Civil Service system can be inflexible. It may take months for a manager to get permission to hire new staff, and several more months to select and bring onboard the chosen candidate. Once hired, a Federal employee enjoys significant protections against dismissal and re-assignment. Agencies often find it difficult to reward top performers with promotions, and substantial pay increases are limited. What's more, Federal employees tend to stay in a position for long periods of time, potentially becoming deeply accustomed to a particular way of doing business and perhaps being tempted to focus more on organizational turf battles than on the societal impact of their work. With few exceptions, government R&D managers must operate within this inflexible personnel system.

In turn, the implications for research results can be significant. The RAGS study, for example, implicitly identified Federal personnel rules as an impediment to the success of government R&D programs when it cited the Defense Advanced Research Projects Agency (DARPA) as a model for a new, energy-focused research agency (National Academy of Sciences, 2007). DARPA is a nonhierarchical agency with an atypical set of human resource authorities. It can quickly hire new staff and pay them significantly more than the typical government worker. In addition, most DARPA program managers are allowed to stay with the program for only a few years, thereby minimizing the extent of bureaucratic insularity. In describing a proposed agency for energy research, the Academies explicitly recommended that it have the same personnel freedoms as DARPA (2007, p. 158) and, indeed, Congress later allowed ARPA-E to hire and pay personnel "without regard to the civil service laws" (42 USC 16538).

The government-wide procurement system also affects the operations of R&D programs. Known as the Federal Acquisition Regulations (FAR), and comprising 4,970 pages of rules and model text,[8] the FAR dictates how Federal agencies procure goods and services (including R&D efforts) from outside vendors. The intent of the FAR is to maintain fairness in competition, prevent fraud and abuse, and ensure that the government pays a reasonable price, but its requirements related to bidding processes, accounting, reporting, conflicts of interest, intellectual property, and information sharing may discourage outsiders from working with the government, especially on R&D projects (Aberman, 2014). As one commentator put it:

> [M]any corporations are refusing to do business with the Government because its regulatory rules are too onerous. The Government is finding that…many corporations will not even accept government dollars to help develop new technologies. (Sidebottom, 2003, p. 87)

In response to this problem, Congress has granted some programs, including DARPA, the authority to engage in "other transactions" that are generally exempt from the FAR (Halchin, 2011). Other transactions[9] authority gives the agency the freedom to ignore the FAR and to negotiate a contract for a technology R&D project with whatever terms and conditions are agreeable to both the agency and the research organization. The National Academies noted the value of the contracting flexibility enjoyed by DARPA and endorsed the concept in its recommendation to form ARPA-E (2007). It is this "other transactions" authority that allows ARPA-E to have the close and entrepreneurial relationship with its research partners that was described in Section 2.4 earlier.

In short, without specific Congressional dispensation, a federally funded R&D program must operate in conformance with the Federal Civil Service requirements for personnel and the Federal Acquisition Regulations for procurement. While there are compelling policy reasons for both systems—related to fairness, transparency, and financial integrity—the effectiveness of Federal R&D programs likely suffers as a consequence.

2.5.3   Inter-agency coordination

In some cases, managers have to contend with R&D initiatives where multiple Federal agencies are directed to work together. It may be that relevant expertise and facilities are scattered across agencies, or that

DOI: 10.1057/9781137542090.0006

more than one agency has a deep interest in the outcome of the research and hence a strong desire to be involved in the conduct of the research. Coordination across agencies may also prevent duplication of effort and save governmental resources. Finally, involvement of multiple agencies in a single initiative may reflect the preferences of Congressional funders, the White House, or key stakeholders outside of government.

Current examples of cross-agency initiatives include the National Nanotechnology Initiative (with 27 agencies and a 2014 budget of $1.7 billion), the Networking and Information Technology R&D Program (13 agencies and a $4.0 billion budget), and the US Global Change Research Program (13 agencies and a $2.7 billion budget) (AAAS, 2013, p. 58). In each of these three examples, there is a multi-agency office responsible for setting strategy, identifying priorities, coordinating research activities, and integrating the results of the many R&D programs within the initiative. In several cases, multiple agencies also share responsibility for individual research programs.

Participation in a multi-agency effort means that administrators must work across organizational lines. Doing so can create challenges related to reporting relationships, control over budgetary resources, and joint decision-making about program activities. In addition, an agency's contribution of resources to a multi-agency initiative likely will create an opportunity cost because fewer resources will then be available for the agency's own programs. White House budget planners acknowledge the dilemma as follows:

> [A]gencies should balance priorities to ensure resources are adequately allocated for agency-specific, mission-driven research, including fundamental research, while focusing resources, where appropriate, on…multi-agency research activities that cannot be addressed effectively by a single agency. (Executive Office of the President, 2014, p. 2)

Given these competing priorities, and in an era of constrained budgetary resources, individual agencies often focus on their own research needs and programs rather than engage deeply in interagency R&D programs (NRC, 2013).

2.5.4   Political oversight

Every government program—be it related to R&D or not—must originate in enacted legislation and be funded through the taxation and appropriation processes. These processes are ultimately under the

DOI: 10.1057/9781137542090.0006

control of elected officials, and in that sense, inherently political. You might not hear a government R&D manager admit (at least in a public forum) that his or her R&D program isn't always operated to maximize its broad societal objectives (as articulated in Section 2.2), but is instead sometimes influenced by politicians who put their own power, stature, and re-election ahead of the program's stated objectives. But the evidence suggests that political oversight of government R&D programs can have a significant impact on their operations.

The importance of political considerations is amplified by another characteristic of government R&D programs: the absence of a single unambiguous guiding objective. While effective program management can be a significant challenge for any R&D manager, at least managers in the private sector usually have a relatively clear objective: maximize shareholder value. In contrast, public sector R&D managers face "diverse, often competing, priorities of bureaucratic superiors, Federal budget controllers, political institutions, and researcher stakeholder groups" (Bozeman & Rogers, 2001, p. 414).

Even when careful cost–benefit analyses can distill an R&D program's impacts into a single metric, perhaps the present value of net social benefits, policy analysts often find that the results are ignored by politicians. While the standard economic logic would suggest that net costs should be minimized or net benefits maximized, politicians often don't see it that way. Much of a politician's job entails ensuring the specific allocation of government resources to eager recipients—usually in his or her home district or state—in a process known as earmarking. Selecting and funding a nationally optimal set of R&D investments may be a lower priority than maximizing the distribution of such resources to constituents (Graetz, 2012). As one knowledgeable participant in a study of Congressional decision-making put it: "For a politician, the costs are the benefits" (Kingdon, 2011, p. 137).

The patterns of political influence over R&D projects were documented by Cohen and Noll who, after case studies of six large Federal programs to commercialize new technologies, concluded that "in the public sector, the ultimate external test of an R&D program is its ability to generate more political support than opposition" (1991, p. 53). The money being spent—on contracts and jobs—in government R&D programsinvariably creates winners and losers who lobby politicians to protect their interests. As programs grow, they often become even more politically salient to members of Congress and "begin to exhibit the familiar characteristics

DOI: 10.1057/9781137542090.0006

of the Federal pork barrel" (p. 74). This pattern leads to a sobering conclusion: weak programs can survive because of intense political support while strong programs may be terminated because of the threat they pose to established political or economic interests.

After a systematic review of Federal science and technology programs operating between 1945 and 2010, Bonvillian similarly concluded that political issues are both exceedingly important to the success of such programs and routinely overlooked during the design and operation of new programs (2011). He goes on to recommend nine design rules that, if followed, can maximize the chances of political success for new programs. For example, new programs can avoid intra-agency rivalry by co-opting existing programs and should be wary of inadvertently growing too large in one geographic location thereby creating political stakeholders who limit program flexibility. The specifics of Bonvillian's nine rules are not especially important to this discussion, but his core conclusion is: political factors can be very important to the success of government R&D programs. If you want to understand how the public sector R&D enterprise operates, it is as important to recognize the influence of politics as it is to understand, for example, market failures and the relationship between R&D and the macroeconomy.

2.6   Model component: outputs

Having used a set of inputs to undertake a series of activities, R&D programs can then be expected to generate some form of outputs. Such outputs comprise the third component of the logic model and can be viewed as integral to the program itself. (More distant results—outside the program itself—are characterized in the fourth and fifth stages of the logic model as, respectively, outcomes and impacts.) The outputs of the typical applied R&D program fall into two categories: the immediate results of the individual R&D projects in the program, and the program's direct effect on the network of researchers, firms, and institutions active in the field.

When it comes to the first, and typically most important, result of an applied R&D project, we usually focus on the new knowledge, insights, and understanding about a technical challenge of interest that has been generated by the project. This body of knowledge—however large or

DOI: 10.1057/9781137542090.0006

small—represents a collection of intellectual property that is the most immediate and proximate output of the investment in R&D. Being intangible, however, such outputs can be hard to measure.

Accordingly, we usually construct one or more proxies for these outputs and then measure the proxy (Jaffe, 2011). Two typical proxies for new knowledge are publications and patents. When it comes to publications, it is common practice to conduct "bibliometric" analysis by counting the number of journal articles authored by the research investigators or which use data generated by the research project. Refinements include considering the "quality" of the journals in which such articles appear, identifying the number of additional publications that subsequently cite these articles, and characterizing networks of subsequent publications to demonstrate how new knowledge from a particular project is combined and re-combined with knowledge from other projects.

Analysis of patent documentation typically uses a similar methodology. Patents directly attributable to a research project are identified, and then subsequent patents that cite the original patent are also identified.[10] Goel and Rich argue that when it comes to applied research, patents are typically viewed as a more appropriate measure of research outputs than are publications (2005). They do note, however, a methodological challenge when it comes to drawing valid inferences from available patent data because the absence of a patent after a project is completed may not be a sign of technical failure, but may instead reflect a firm's business decision to protect its new technology as a trade secret, rather than with a patent.

Beyond a review of the patents and/or publications that follow from an R&D investment, we can also characterize outputs in terms of the practical significance of the research results: whether it represents technical success, or potentially even a technical breakthrough, or whether it is trivial and unimportant, or at worst a technical failure.

It is usually unhelpful to think simplistically about whether a project has succeeded or failed; instead, a more nuanced appreciation of the continuum of potential results is needed. At one end of the continuum are projects that are indeed genuine failures. Whether because of an ill-conceived research design, inadequate facilities, an insufficiently skilled research team, failure of project partners to perform as agreed, weak project or human resource management, or excessive political interference, it is possible to come to the end of a research project and have virtually nothing to show for it. In a more positive vein, a project

DOI: 10.1057/9781137542090.0006

may be well designed and executed, but still fail to achieve its technical objectives. Perhaps the engineering challenges were just too great, or maybe the underlying basic science was discovered to operate in unexpected ways that impede technical success. In such cases, however, valuable knowledge may still be gained about what does not work, thereby informing future research initiatives that can then focus on other areas of inquiry.

As regards projects that at least partially achieve their technical objectives, it is important to remember that especially in the field of applied R&D, we are usually interested in both the cost and performance of newly developed technologies. For example, a new computer chip with a processing speed twice its nearest competitor might constitute a technical breakthrough, but if its cost is also twice as high, then the R&D project probably won't be considered an unqualified success. Similarly, if researchers discover a way to generate new forms of electric energy at a bench scale, but are unable, or not tasked or funded, to develop a manufacturing process that can cost-effectively deliver the technology on a mass scale, then their research result remains incomplete. Given the accretive nature of knowledge, and the complexity of new technologies, however, even a project that itself doesn't create radical new innovations may ultimately serve as a steppingstone toward such innovations. Finally, at the end of the "failure versus success" continuum are projects that fully meet their objectives by developing new technologies with cost and performance characteristics unambiguously superior to those of existing technologies.

Beyond the intellectual property and technologies that emerge from publicly sponsored R&D programs, another key output of such programs is their contribution to the formation and maintenance of networks between and among scientists, engineers, and entrepreneurs working in the commercial, public, and academic sectors. There is a robust literature that suggests the depth and size of such networks is a key driver of the degree of innovation that takes place over time (Vonortas, 2013).

Outputs from an R&D program related to changes in research networks might include the number of previously unpublished or unfunded researchers added to the community by virtue of their participation in a sponsored project, joint publications or patent applications between members of the research team or between a team member and someone outside the team, the number and depth of newly created relationships between institutions that had not previously worked together,

DOI: 10.1057/9781137542090.0006

and the duration (one-off or long-term) of the personal and institutional connections fostered by the research program.

2.7 Mediating factors: conditions faced by new technologies

After reviewing the history of technical progress between 1760 and 1940, Joseph Schumpeter coined the phrase "creative destruction" to describe the processes by which successful technological innovations diffuse through the economy (1943, p. 83). Although dated, Schumpeter's turn of phrase still captures the important point that two phenomena—creation and destruction—must both be present for significant technological change to occur. Inventors, innovators, and entrepreneurs must first "create" new technologies that offer the prospect of increased value for potential users (who could be firms or individuals). Users must then decide or be compelled to abandon existing technologies and ways of doing business (hence the "destruction" metaphor) and adopt the new technology.

Scholars have spent several decades working to understand how new technologies make their way into everyday use by firms and consumers (Rogers, 1995; Reinganum, 1989; David, 1985; Kamien & Schwartz, 1982; Rosenberg, 1972; Mansfield, 1961). A review of this literature and a complete treatment of all the factors that drive the process of technology diffusion is beyond the scope of this analysis. A general understanding of the topic, however, is a prerequisite to an appreciation of the potential outcomes and impacts of a publicly supported R&D program. In particular, it is helpful to understand the basic market dynamics that affect the rate of technology adoption, as well as the policy-driven impediments and incentives that can affect this process.

2.7.1 The market environment for new technologies

To understand the market environment for new technologies, we can start with the decision calculus of an individual or firm considering the adoption of a new technology. We can then step back to a market-wide perspective to think about the implications of these micro-level decisions for the diffusion, or market penetration, of a new technology.

When a potential user becomes aware of a new technology, it faces a decision: adopt it or postpone consideration of it until later. Hall (2005)

DOI: 10.1057/9781137542090.0006

and Tassey (1997) both describe three economic factors that influence this decision: benefits, costs, and uncertainties.

First, the potential adopter must assess the likely incremental benefits of the new technology over those of the technology currently in place. Such benefits may arise in the form of lower operating costs, higher quality outputs, or new products and services more highly valued by buyers. When it comes to smartphones, for example, innovation is driven by product differentiation that focuses on new features that users see as valuable (e.g., GPS tracking, voice recognition, durable touchscreens, a large inventory of applications). On the other hand, the rapid diffusion of fracking technologies for natural gas production is driven primarily by the cost advantage and productivity increases associated with the new technology, rather than any material difference in the quality of the methane gas being produced.

In some cases, network effects may also be important to this benefit calculation. Network effects can arise when the value of a new product depends in part on how many other users already exist. The value, for example, of an internet connection depends, in large measure, on how many of one's friends and colleagues (or, for a firm, how many suppliers and customers) have access to the Internet. Another form of network effect can arise when a new technology requires complementary products or services in order to create value for a customer. Even though automakers may be able to design and produce cars fueled by natural gas, for example, such vehicles are unlikely to enjoy widespread adoption unless there is a network of natural gas refueling stations.

Weighed against the potential benefits of the new technology are the costs of adopting it. Acquisition and operating costs are one part of the calculus, but so too are the costs of necessary complements that must be acquired, as well as the transition cost of integrating the new technology into current operations. Because it likely operates differently than the current technology, and may require modifications to existing business processes, adopters of a new technology often must incur the cost of climbing a learning curve to realize its full value. If these transition costs are large, users of current technologies may perceive they are "locked-in" and unable to adopt a new technology even though it may, from a narrow view, offer positive net benefits.

Finally, in addition to the estimated costs and benefits of the new technology, potential users must contend with significant uncertainties. How will the new technology actually perform? As advertised, or will hidden

DOI: 10.1057/9781137542090.0006

problems (and costs) emerge? Will complements be available, at what price? What about service and maintenance support? How difficult will it really be to integrate the technology into current ways of doing business? Will competitors adopt the technology? How will customers react to the changes associated with the technology? Such uncertainties create risk and, depending on the assessment of risk (and the decision maker's level of risk-aversion), may slow the diffusion process, as decision-makers wait for uncertainty to diminish before making a decision about adopting the new technology.

Having considered the calculus of individual decision-makers, we can now step back and take a market-wide perspective. The market penetration of a new technology depends on the choices of a heterogeneous set of firms and consumers who will likely perceive different costs, benefits, and uncertainties from a potential new technology and who will make choices about whether to adopt it at different points in time. It is the aggregation of these individual decisions that drive the overall diffusion of a new technology in the market.

It is important to remember that this decision calculus is almost always based on a private, rather than a social, perspective. Accordingly, if adoption of a new technology would create either positive or negative externalities (e.g., decreased or increased greenhouse gas emissions), those externalities are unlikely to drive the aggregate technology diffusion process.[11] An exception might exist if large numbers of potential new technology users are motivated by social costs and benefits, perhaps because of corporate social responsibility policies or concerns about brand image among retail customers.

If a new product requires more than a few potential users to abandon a substantial base of existing technology and to incur high transition costs, or if needed complementary products and services are scarce, we can't expect rapid market penetration, especially if the cost advantage of the new technology is slight or if its incremental benefits are limited. Similarly, if uncertainty about the ultimate value of a new technology is pervasive, diffusion will proceed at a slower pace than it otherwise would. Finally, providers of existing technologies are unlikely to simply surrender their markets to the new technologies; instead, these incumbents will almost certainly cut prices and endeavor to enhance the existing technologies to improve the value they deliver to customers. To the degree they succeed in doing so, diffusion of the new technology will be slowed.

DOI: 10.1057/9781137542090.0006

By way of example, Moniz explains the slow market penetration of renewable energy by first noting that the energy system is a "business with high barriers to displacement of incumbents" (2012, p. 83). To start, electricity is a commodity, meaning that electricity generated from a coal plant is the same as electricity from a wind farm, thus removing a potential quality difference from the decision calculus. In addition, existing and new power-generation assets are very costly, making utilities reluctant both to abandon existing capital investments in fossil-fuel-fired power plants and to make large new investments in renewable energy generation. Finally, an important (and expensive) complementary good—transmission lines—often don't exist in close proximity to sources of renewable energy. These factors lead Moniz to conclude that "early adoption of energy technology is highly sensitive to policies and subsidies that help create the initial market" (2012, p. 83).

2.7.2   The policy environment for new technologies

As Moniz points out, technology diffusion is not just a reflection of competitive markets operating free of government intervention, but also of policies that alter the costs, benefits, and uncertainties associated with new technologies. And, of course, such policies don't arise in a vacuum; instead, they are the result of a political process. These two topics—policy and politics—are addressed briefly below.

Policy can affect technology diffusion either by changing the relative costs of new and/or existing technologies or by using regulation to change the "rules of the game" in a way that advantages one technology over another. Government can change relative costs with taxes and subsidies. The Federal government, for example, has provided tax credits for the construction of solar facilities and for the production of wind energy. Its temporary "Cash for Clunkers" program provided cash rebates for customers who bought high mileage cars and scrapped older, lower-mileage cars. Conversely, policies can make some technologies more expensive. The "Gas Guzzler Tax," for example, imposes a progressively higher sales tax on low mileage cars, reaching a maximum tax of $7,700 for a car that gets less than 12.5 miles per gallon (U.S. EPA, 2012).

Government can also use its regulatory authority to affect the diffusion of new technologies. Recently enacted changes in the corporate average fuel economy standards, for example, require automakers to increase the efficiency of new passenger vehicles from a current level of 33.6 miles per gallon to 55.3 miles per gallon in 2025 (C2ES, 2014). Similarly, 30 states

DOI: 10.1057/9781137542090.0006

have imposed Renewable Portfolio Standards (RPS) that require utilities to source a specified fraction of their electricity from renewable sources (EIA, 2012). On the other hand, RPS requirements notwithstanding, most states require utilities to the deliver the lowest-cost energy available which creates a regulatory bias toward certain types of power generation (Moniz, 2012).

The tax, subsidy, and regulatory policies described earlier do not originate only from the decisions of policymakers anxious to maximize social welfare writ large. They also reflect a political process. To the extent that firms have political power, technology incumbents may endeavor to block the diffusion of new technologies while innovators may seek government support to facilitate their success. This phenomenon has been summarized as follows:

> [P]olitics are often neglected by…researchers, who tend to assume widespread support for progress in science and technology. In [several] cases, the losing interest groups were able to convince politicians to block, slow, or alter government support for scientific and technical progress.… Exactly who are these "losers"? Depending on the form it takes and the economic and political environment in which it appears, technological change can threaten labor, corporations, consumers, governments, and so on.… Resisters can organize and use their financial or electoral clout to influence government to slow technological change via a range of mechanisms: taxes, tariffs, antitrust legislation, licensing, standards setting and regulations, manipulation of guidelines for research, and so on. (Sapolsky & Taylor, 2011, pp. 33–34)

It is not only technology incumbents who engage the political process to protect their business interests. Innovators are often seen to engage in similar tactics. Press reports suggest that wind power, for example, enjoys considerable political support, even from politicians who deny the existence of climate change, because it brings economic development to their home districts (The Economist, 2013). After enticing five states to compete for the opportunity, Tesla Motors selected Nevada as the home of its new large lithium-ion battery factory, after Nevada politicians decided to grant it $1.25 billion in tax breaks over 20 years (Pyper, 2014).

2.8   Model component: outcomes

After an R&D program creates immediate outputs in the form of new intellectual property, technical project results, and new linkages in the network of innovators, we might expect those outputs (which are tightly

DOI: 10.1057/9781137542090.0006

connected to the program itself) to lead to other downstream results that go beyond the R&D program and which are generally not subject to the direct control or influence of program managers. In the parlance of logic models, such results are referred to as program outcomes (i.e., the fourth step in the model). Given our focus here on nondefense applied R&D, a key outcome of interest is the degree to which new technologies succeed in the marketplace. After all, technologies that are not deployed cannot be expected have a significant impact on economic prosperity, national competitiveness, job creation, or environmental protection.

When thinking about the concept of the market success of a new technology, it can be helpful to consider three interrelated factors. The first is the extent to which the new technology ultimately penetrates the market. Until recently, for example, almost every four-wheeled vehicle on the road relied on the internal combustion engine (ICE) for power and thus the ICE enjoyed a market penetration rate of nearly 100 percent. Conversely, while very popular, Apple's iPhone has penetrated only 42 percent of the current smart phone market (MobiLens, 2014).

The second aspect of market success is the speed with which a new technology penetrates the market. In the field of consumer electronics, for example, the video cassette recorder reached 70 percent market penetration[12] in ten years, while the CD player took 14 years and cell phones took 19 years to achieve the same level of market penetration (Schilling, 2010, p. 58). The faster a technology penetrates the market, the more quickly it will generate financial returns to its developers and, if there are spillover social benefits, the sooner society will experience those benefits.

A final aspect of market success is its duration. Most products and services face ongoing competitive pressures as other firms and entrepreneurs endeavor to replace them with new offerings that, by virtue of price and/or performance, are more attractive to buyers. Even a moment's reflection brings to mind a long list of technologies that once enjoyed dominant market positions, but are now little more than historical relics: horse-drawn carriages, steam locomotives, mainframe computers reliant on punch cards, subway tokens, film-based cameras, vinyl records, and so on. New technologies that enjoy a short period of market success before being eclipsed by a competing technology are likely to create a lower return to the investment in R&D than technologies that experience long-term, enduring success.

DOI: 10.1057/9781137542090.0006

Figure 2.4 integrates these three factors—maximum market penetration, speed of market penetration, and duration of market success—in a two-dimensional framework. Three fictional technologies are illustrated.[13] Technology A penetrates the market slowly but enjoys a long period of near-complete market dominance before beginning to lose market share. Technology B penetrates the market more quickly and also experiences a stable market share, but it never captures more than 50 percent of the market. Technology C succeeds in quickly taking a majority of its market, but is able to hold that position only for a short period before ultimately being forced out of the market.

Accordingly, when considering the technical outcomes of a public sector R&D program, it is important to go beyond a simple yes/no answer to the question of whether a new technology that resulted from the R&D program has been successful in the market. The extent, speed, and duration of market penetration—factors illustrated in Figure 2.4—tell a more complete story about the outcomes of the program and, in turn, allow for a more careful analysis of the returns to the investment in the R&D program.

A second potential outcome of an R&D program is an increase in the innovative productivity of the networks among researchers in government, academia, and the private sector. To clarify: changes in the number of researchers in the network, in the number of connections in

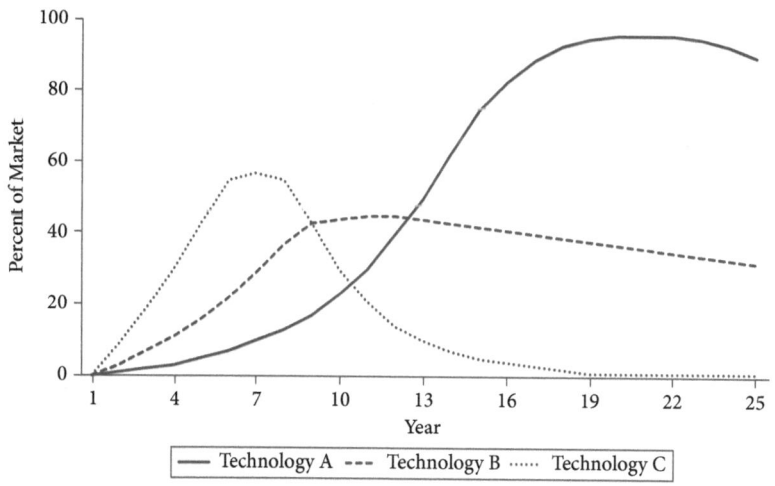

FIGURE 2.4 *Illustrative market penetration pathways for new technologies*

DOI: 10.1057/9781137542090.0006

the network, and in the density of the network would constitute program *outputs*. Only if such changes result in increased levels of innovation from the community would it then be appropriate to credit the R&D program with producing a meaningful *outcome*.

2.9   Model component: impacts

Having carefully delineated the outcomes of a particular R&D program (e.g., the market penetration of a new technology or the enhancement of networks in the research community), we are ready to consider the final component in the logic model: the ultimate impacts of the program. Before doing so, however, we need to tackle one additional analytic issue. To this point, we have been looking only at the R&D program itself, and at the outputs and outcomes it has created. When we endeavor to characterize a program's long-term impacts, however, we are asking a broader question: has the program made a practical difference in real world conditions?

The word "difference" is a tipoff about this analytic issue; to understand program impacts, we need to understand what would have happened in the absence of the program. In short, if we want to causally attribute specific impacts to a particular R&D program, then we need to understand *two* states of the world. The first, observable, state of the world includes the program and all of its outputs and outcomes. The second, referred to as the counterfactual, describes a world in which the program does not exist. In turn, it is the incremental gap between these two states of the world that defines the ultimate impacts of a program. If such a gap can be discerned, then the program is said to exhibit "additionality."

Consider a government R&D program that develops a new technology which, had the program not existed, would have instead been developed by the private sector on roughly the same schedule at about the same cost. In this case then, the government program would have had no meaningful impact on society, despite having created both outputs and outcomes. A range of methodologies exist for specifying a counterfactual situation and for computing incremental impacts. A more general treatment of these topics is beyond the scope of this book, but several potentially helpful references are cited later in Section 2.10 (Program Evaluation).

In many ways, understanding the final component of the logic model (i.e., its impacts) takes us back to its beginnings: what objectives

DOI: 10.1057/9781137542090.0006

did we have for the R&D program in the first place, and what did we hope to accomplish by funding and operating it? If we were trying to correct market failures that reduce the level of R&D investment below some social optimum, then the primary impact of interest would be an increase in the level of R&D that in turn spurs increased innovation, new products with higher performance and lower costs, new services that create value for consumers and firms, or higher productivity from labor or from capital assets. If the focus had been on enhancing national competitiveness, then relevant impacts could include faster economic growth, lower unemployment, a more skilled workforce, or development of new industries. Finally, if we intended the R&D program to foster new domestic, environmentally friendly, energy technologies, then the impacts of interest would likely include incremental reductions in emissions, improvements in environmental conditions, fewer energy supply disruptions and price spikes, and a reduction in military costs (both human and monetary) associated with securing global energy supply chains.

When it comes to the downstream impacts of program outcomes related to building networks among researchers and entrepreneurs, we are reminded about the limitation of a linear logic model. Stronger research networks are not an end in themselves. We are interested in a stronger network because of its potential to foster more innovation, economic growth, and so forth. Accordingly, for program impacts that are themselves the inputs to another cycle of R&D initiatives, we need to think in a nonlinear fashion and consider the feedback loops that will be important not only in the program under consideration, but in future research endeavors as well (Jordan, 2013).

In addition to its intended impacts, an R&D program may also create impacts that are unrelated to its initial objectives. In what is now a classic example, the roots of the modern Internet lie in ARPAnet, a system created to allow Defense Department researchers to share files among different geographic locations. These researchers, at least initially, did not intend to create the Internet, but instead were simply looking for a way to make their daily interactions more productive. Nonetheless, ARPAnet and its downstream effects on the Internet can fairly be called an "impact" of DoD's R&D programs.

Finally, it is important to note that not all impacts of a government R&D program are beneficial. I am not focused here on those R&D programs that fail to protect the health and safety of human subjects,

DOI: 10.1057/9781137542090.0006

but instead on the diversion of resources to the R&D program and on the opportunity costs that consequently arise. Had funds not been spent on an R&D program, they would have been spent on another government program or not raised from taxpayers in the first place. Either way, those resources would likely have generated some private and/or social benefits, the loss of which, while very hard to estimate, conceptually are an impact of the R&D program.

2.10   Model component: program evaluation

To re-cap, the proposed framework for thinking holistically about the public sector R&D enterprise comprises five core components (inputs, activities, outputs, outcomes, and impacts), as well as three sets of mediating factors (policy objectives, institutional conditions, and the environment faced by new technologies). While this framework can be used prospectively to inform the design of a new R&D program, it can also be used retrospectively to evaluate an existing program. Program evaluation allows public administrators to extract lessons learned from an existing R&D program that can then be applied in several ways (Linquiti, 2012a). First, if the studied program is still underway, the results of the evaluation can be used to improve its performance. Second, there exists a normative view that when taxpayers' money is involved, public administrators ought to be held accountable for their use of those funds and the results that are obtained (Link & Scott, 2011) and a program evaluation can provide a means of assuring accountability. Finally, the findings of a program evaluation about what works, and what does not, can provide helpful evidence to inform the design of other, nascent R&D programs.

When it comes to generic approaches to program evaluation (i.e., not specific to R&D programs), the Government Accountability Office (GAO) has published detailed guidelines for its evaluators to use as they assess government programs (GAO, 2012).[14] The GAO approach to program evaluation entails five steps, briefly summarized as follows:

1   Careful delineation of program goals and objectives (i.e., the initial rationale for the program) and the construction of a program-specific logic model.
2   Development of research questions for each component of the logic model which, when answered, will provide a complete picture of how that component has operated in the studied program.

3 Selection of a research design capable of yielding reliable answers to the research questions posed in the second step and that can also be used to establish a reasonable counterfactual to allow calculation of incremental program impacts and make inferences about causality.

4 Specification of data sources and collection methods, which could range from interviews of program stakeholders to extracts of administrative program records to collection of market data on prices and quantities sold for new and existing technologies.

5 Development of a data analysis plan that specifies the ways in which qualitative data will be evaluated, what quantitative techniques will be used, and how the qualitative and quantitative analyses will be integrated in a way to ensure both validity and reliability of the results.

After the program evaluation is designed, the research is then conducted. Data are collected and analyzed in accordance with the research plan, and then the program evaluation is written. Depending on the audience, different versions of the evaluation may be created. GAO, for example, creates a single-page summary of each of its evaluations, as well as a much more detailed report for program administrators responsible for putting specific recommendations and findings into practice.

Other resources are available to assist the analyst specifically interested in evaluating an R&D program. Perhaps the broadest treatment of the topic to date is the previously cited volume edited by Link and Vonortas (2013); other references tend to focus more narrowly on the microeconomic questions of costs, benefits, and returns on investment, but can still be quite helpful in tackling methodological issues such as the development of a counterfactual, the attribution of causality, the diffusion of new technologies in the market, and the monetization of benefits (Ruegg & Feller, 2003; Powell, 2006; Ruegg & Jordan, 2010; Link & Scott, 2011; Link & Scott, 2012; Tassey, 2003).

Notes

1 In a parallel process, the Federal Reserve uses "monetary policy" to achieve similar ends by managing interest rates and the money supply.

2 Recall that, as mentioned previously, Federal spending on applied, nondefense, R&D is about $30 billion per year.

DOI: 10.1057/9781137542090.0006

3   Johnson cites scientist Stuart Kauffman as the first to use the phrase "adjacent possible." Kauffman did so in the context of biochemistry, rather than technological innovation.

4   Incomplete or clearly deficient proposals are usually not subjected to intensive review.

5   The legal basis for ARPA-E's distinctive approach is explained in more detail in Section 2.5.2.

6   Broad requirements applicable to all Federal agencies, rather than specific agencies and programs, are discussed in the next section.

7   More information about ATP can be found in Link and Scott (2011).

8   On the basis of information from the Government Printing Office website at bookstore.gpo.gov, as of November 2, 2014.

9   The adjective "other" is used to describe transactions that are not grants, cooperative agreements, or traditional Federal contracts.

10 This discussion highlights the fuzziness of the borders between the components of a logic model. Strictly speaking, the set of publications and patents created directly by research program participants or using program data would constitute a program "output" while subsequent publications and patents by others would constitute a program "outcome."

11 If, as a matter of policy, such externalities have been internalized (perhaps by a "price on carbon"), then they will indeed influence the diffusion process. Of course, if they have been internalized, then technically speaking, they are no longer externalities.

12 Measured as the percentage of US households owning the technology.

13 These three technologies are assumed to not be competitors in the same market.

14 Another good summary of the field of program evaluation writ large is Wholey, Hatry, and Newcomer (2010).

3

R&D Portfolio Valuation and Formation

Abstract: *Chapter 3 confronts the challenge of predicting the return on R&D investments before they are made. It critiques traditional government methods that apply discounted cash flow analysis to individual R&D projects in isolation. Option theory and portfolio theory are used to identify potential enhancements to traditional methods. Chapter 3 also demonstrates that focusing exclusively on potential returns while ignoring risk is a fundamentally unsound practice. Methods often used by the private sector to form, value, and manage portfolios of R&D projects are summarized, including the "bucket" method, quantitative multi-objective methods, static choice methods, and dynamic management methods. While a portfolio perspective is rarely taken within government R&D programs, a few examples are identified and characterized.*

Keywords: dynamic R&D management; government R&D; NRC Prospective Evaluation Study; real options; R&D portfolios; R&D risks

Linquiti, Peter D. *The Public Sector R&D Enterprise: A New Approach to Portfolio Valuation.* New York: Palgrave Macmillan, 2015. DOI: 10.1057/9781137542090.0007.

Having described in broad terms how the Federal government typically operates its R&D programs, I now turn to a much more specific topic: the selection of R&D projects within a program.[1] In the language of logic modeling, my focus here is on the "Activities" component of the model, in particular the second and third activities described in Section 2.4 (i.e., project evaluation and portfolio formation).

This analysis is intended to help policymakers more accurately value prospective investments in R&D projects and portfolios, as well as to better understand the risks associated with such investments.[2] I start with a theoretical critique of the methodology developed by the National Research Council (NRC) for use by Federal R&D managers and offer some proposed enhancements to that methodology. I then quickly review some typical practices, seen in both the commercial and public sectors, for forming and valuing portfolios of R&D projects. On the basis of these analyses, I turn in Chapter 4 to the demonstration of an alternative to the NRC approach.

3.1   National Research Council method for prospective evaluation

When public sector R&D is prospectively subjected to quantitative economic analysis, the usual approach is borrowed from private capital budgeting methods and analyzes costs and benefits in a discounted cash flow framework (Vonortas & Desai, 2007). An exemplar of this approach is the work of the NRC which, at the behest of Congressional appropriators (U.S. Congress, 1999; 2002), developed and applied a methodology for prospectively estimating the expected benefits of R&D projects conducted by the US Department of Energy (DOE) (NRC, 2005; 2007). This R&D is intended to foster new technologies for energy production and to improve energy efficiency, thereby mitigating climate change, improving energy security, and providing a source of national competitive advantage.

For example, one component of the 2007 NRC study estimated the prospective benefits of a portfolio of 22 R&D projects sponsored by DOE's Chemical Industrial Technologies Program. NRC's methodology combined a decision-tree analysis with a traditional discounted cash flow (DT/DCF) framework. NRC estimated the probabilities of technical and commercial success for each R&D project and then applied a forecast of the likely energy cost savings through 2030 that would result if the new

DOI: 10.1057/9781137542090.0007

technology generated by the project were to be deployed.[3] The result was the expected value of the benefits of each R&D project. Improvements in national energy security and reductions in pollutant emissions were also assessed, but neither was monetized. Interdependencies among the R&D projects in the portfolio were not analyzed.

While an important step forward, the NRC methodology suffers three important shortcomings. First, it uses a point estimate, rather than a range, for the value of annual energy savings that might result from each new technology. A more realistic treatment of the savings would incorporate a probability distribution for such savings and then explicitly link the probability of commercial success to the size of the savings. Traditional cost-benefit analysis techniques that ignore the opportunity to change course in the future as uncertainty is reduced may systematically undervalue the investment (Copeland & Antikarov, 2003; Shockley, 2007; Triantis, 2003). A good example of the problem with NRC's approach is its treatment of two R&D projects in the Chemical Industries Program for which forecast energy savings are negative (i.e., efficiency would actually decrease and energy costs would rise); that notwithstanding, the NRC investigators assigned both projects a nonzero probability of commercial success. The implausibility of such an outcome highlights the danger of a disconnect between estimates of market value and the likelihood of market success.

A second shortcoming in the NRC study was the failure to analyze interdependencies among R&D projects in the portfolio. While the study methodically estimated the probability of market and technical success for each project, it did so by treating each project as a separate undertaking. In reality, however, these probabilities of success are almost certainly not independent. Two R&D projects, for example, might be attacking the same technological challenge in fundamentally different ways and the success of one project might indicate that the other project is less likely to succeed. Or two projects might be focused on reducing the consumption of electricity by industry; if the price of electricity rises, then the value of both projects will rise as a result. As demonstrated later in this chapter, ignoring these potential interdependencies among the projects in an R&D portfolio can significantly distort the estimated risk associated with the R&D investment.

To be fair, the potential for interaction among projects was noted by the NRC investigators, but not analyzed:

> [M]ore work is required to describe a method for estimating the benefits of DOE's overall portfolio of separate programs or subprojects.... The committee

DOI: 10.1057/9781137542090.0007

should undertake the following activities in Phase Three…determine how the benefits methodology can be applied for portfolio analysis and evaluate a portfolio. (NRC, 2007, pp. 4, 60)

Ironically, Congress terminated the NRC study and declined to fund the aforementioned Phase Three, saying "the [House Appropriations] Committee believes that there is no further benefit to be gained from the study effort on Prospective Benefits of DOE's Applied Energy R&D Programs" (U.S. Congress, 2007a, p. 64).

A third shortcoming of the NRC study is that, while it provides benefit estimates for discount rates of 3 and 7 percent, it offers little guidance on how to interpret the two estimates.[4] According to the Office of Management and Budget (OMB), 7 percent is the real, inflation-adjusted, "average before-tax rate of return to private capital in the U.S. economy" and "approximates the opportunity cost of capital" while 3 percent represents the "social rate of time preference" based on the "real rate of return on long-term government debt" (OMB, 2003, p. 33). One view of the difference between OMB's rates of 3 and 7 percent is that it constitutes a risk premium. Long-term US government debt (the source of the 3 percent rate) is generally considered to be risk-free while the average cost of capital in the economy (the source of the 7 percent rate) reflects a premium for risks faced by investors holding a broad portfolio of securities.

The use of a single, risk-free, discount rate (i.e., 3%) might be defended with the argument that public sector program administrators either do not or should not exhibit risk aversion. The use of a single, risk-adjusted, discount rate (i.e., 7%) that includes a risk premium is, however, difficult to justify. Because risk is likely to vary over the life of the R&D project, as well across its component costs and benefits, the risk premium should be adjusted to fit the particular cost component and point in time.[5] When it applied a discount rate of 7 percent, the NRC panel implicitly attached a constant risk premium of 4 percent to all costs and benefits. A simple example demonstrates the problem with this approach: the NRC study discounts short-term R&D costs at the same rate as the long-term commercial value of the research results. The former are typically codified in agreements among R&D sponsors and are almost certain to be incurred as planned while the latter results from a highly uncertain interaction of the price and performance of the sponsored and competing technologies, energy prices over time, the regulatory environment, and general macroeconomic conditions. The risks and uncertainties are

DOI: 10.1057/9781137542090.0007

clearly different between the two cases and a compelling argument can be made that the discount rate (and risk premium) should vary as well.

3.2   Potential enhancements to the standard DT/DCF approach

While it was an important step forward, the NRC methodology—with its focus on decision trees and discounted cash flows—could be improved. Accordingly, I propose three enhancements that would address most of the shortcomings identified earlier.

3.2.1   Enhancement #1: treat R&D projects as real options

The similarities between financial and real assets have attracted considerable scholarly attention, especially in regard to the managerial flexibility associated with real assets such as business ventures, technical knowledge, and natural resources. Stewart Myers appears to have been the first to refer to such flexibility as a real option (1977); other work in this field includes Kester (1984), Dixit and Pindyck (1994), Trigeorgis (1996), and Ram and Goel (2009).

Financial and real options are similar in that both give the option holder the right, but not the obligation, to take an action at a future date. Financial options are linked to financial securities while a real option is associated with a real underlying asset, such as oil reserves, or in the case of R&D, knowledge and information. Typically, the right is the option owner's opportunity to buy (i.e., to "call") or to sell (i.e., to "put") the underlying asset to another party (the option writer). Real options are likely to be valuable when future outcomes are uncertain, the uncertainty is likely to be reduced over time, the flexibility exists to take action in the future as the uncertainty is resolved, and the action can reduce costs or increase benefits when it is taken (Triantis, 2003).

The Science of Science Policy Task Group specifically identified real option valuation as fully "relevant" to R&D investments, but not yet a fully "mature" tool for use by policymakers (ITG, 2008, p. 26). This analysis aims to be a step forward in that maturation process.

Real option analysis has been applied in many private sector contexts including corporate R&D, oil and gas exploration projects, mergers and

DOI: 10.1057/9781137542090.0007

acquisitions, and real estate development projects (Triantis, 2001; Shockley, 2007). There also exists a modest, but growing, body of literature related to public sector applications. Real option analysis has been applied to public sector technology and R&D initiatives in the areas of hydrogen and fuel cell technologies (Mahnovski, 2007), intelligent transportation systems (de Neufville, Hodota, Sussman, & Scholtes, 2008), the Department of Energy's science programs (Vonortas & Desai, 2007), supersonic transport (Vonortas & Hertzfeld, 1998), and renewable energy (Davis & Owens, 2003).

When it comes to framing an R&D project as an option, the appropriate analogy is usually to a call option. The R&D sponsor, by virtue of its investment in R&D, may create an asset of some value (i.e., the knowledge gained by execution of the research). In the same way that the holder of an option on a stock is uncertain about future stock prices, the R&D sponsor is uncertain about the ultimate market value of the knowledge generated by the project. At the end of the project, the option owner has the right, but not the obligation, to use the new knowledge if it proves sufficiently valuable. In the parlance of options analysis, the cost of the R&D project is equivalent to the option purchase price and the cost of converting its results into a new operational process or commercialized product is equivalent to the option exercise (or strike) price.

Having framed the real option, the next task is to value it. It is instructive to begin by considering the underlying mechanism that drives the payoff to an option. Figure 3.1 depicts the payoff for a real R&D option. The horizontal axis measures the market value (V) of the result of the R&D effort. The deployment of the new technology has some cost X (i.e., its exercise, or strike, price), and the option owner is assumed to deploy the technology (i.e., exercise the option) only if its commercial value exceeds the deployment cost (i.e., if V>X and the option is "in the money"). The payoff to the option [max(0,V-X)], measured on the vertical axis, is zero if the market value is below the deployment cost and rises on a one-for-one basis with the market value at points above the deployment cost (hence the 45° line). Observing the option payoff graph shown in Figure 3.1, we can see that the "downside" risk of an option is truncated; if the underlying asset value falls below the strike price, the investor's loss does not fall commensurately. The most that the investor will lose is the initial price of purchasing the option. Conversely, the investor's "upside" is driven the shape of the probability distribution that characterizes the value of the underlying asset (i.e., the results of the R&D) for values greater than X. In either case, however, the probability

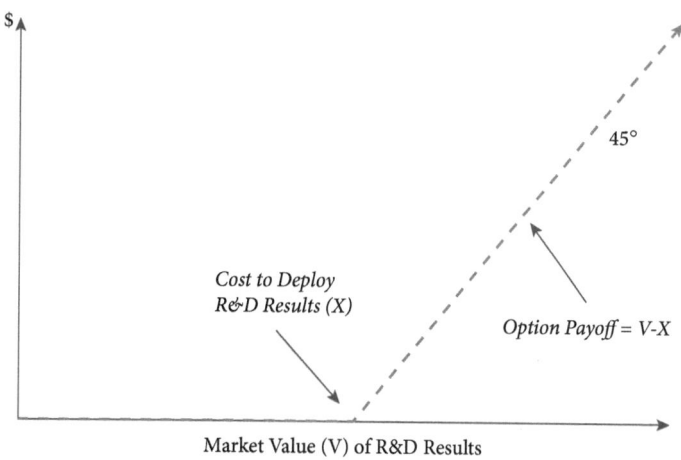

FIGURE 3.1   *Payoff to an R&D option*
Source: Linquiti (2012b).

distribution for the option's value is contingent on the probability distribution of the underlying asset.

Figure 3.2 supposes an R&D project with a normally distributed payoff that has a mean (μ) of \$100 and a standard deviation (σ) of \$10. In this simple case, the cost of deploying the R&D result is assumed to be \$110 and the option thus has value if the underlying value exceeds \$110 (i.e., the option only has value if the payoff is in the shaded area to the right of the distribution). Given the nature of the normal distribution, there is a 15.8 percent chance of a nonzero payoff and, as is clear from visual inspection, the distribution of the payoff is not normally distributed. The mean payoff (i.e., the mean of the shaded area) is \$0.83, and the standard deviation is \$2.61. Both parameters are radically different from the original distribution (where $\mu = 100$, $\sigma = 10$).

There are many option valuation methodologies, but virtually all of them depend on one of two theoretical constructs. The first, referred to as the replicating portfolio approach, is built on the idea that the value of an option equals the market price of a portfolio of securities that replicates the risk and cash flow of the payoffs of holding the option. The replicating portfolio for a call option on a traded stock can be formed by buying shares in the underlying asset (i.e., the stock) and borrowing money for a term equal to the duration of the option. The mix of shares and borrowing can be set to ensure that, for all possible future stock prices, the replicating

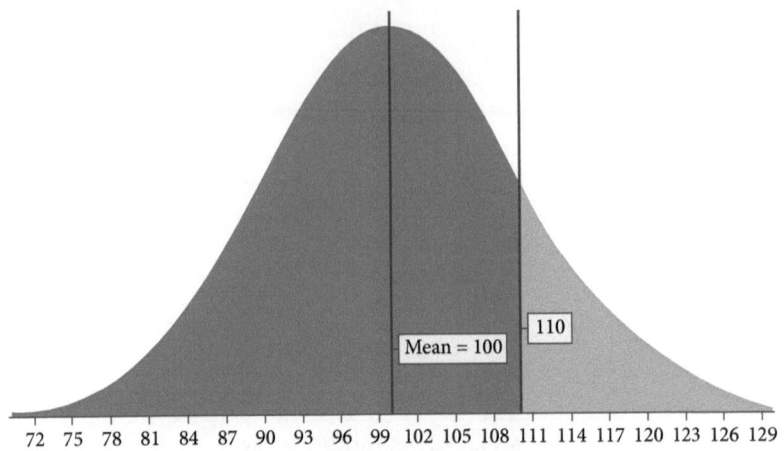

72 75 78 81 84 87 90 93 96 99 102 105 108 111 114 117 120 123 126 129

FIGURE 3.2   *Underlying asset value and option payoff*

portfolio has a payoff identical to that of the option. If the option weren't valued at the computed price of this portfolio, then an arbitrage opportunity would exist and, assuming an efficient market, it would quickly be exploited by market participants, thereby equating the option value with the value of the replicating portfolio (Brealey & Myers, 2003; Hull, 2009).

The second construct for valuing options is referred to as risk-neutral valuation. It produces the same result as the no-arbitrage, replicating portfolio approach, but does not require the underlying asset to be traded in the market, making it especially useful for valuing real options. An R&D project is seldom traded in the marketplace, so there is no ready source of data about its market value.[6] Instead, the analyst must estimate a value as if it were traded (Brealey & Myers, 2003). Risk-neutral valuation builds on the insight gleaned from the no-arbitrage method that risk preferences do not affect option values (as long as efficient markets extinguish arbitrage opportunities).

If risk preferences are irrelevant, then options can be valued in a risk-neutral world where investments in the underlying asset do not command a risk premium and instead grow at a risk-free rate, albeit with the same variability as in the "real," as opposed to risk-neutral, world. At the time of option expiration, any potential payoffs (i.e., the asset value minus the exercise price) can be discounted back to the current period at a risk-free rate. The probability-weighted expected value of the payoffs is equal to the option value.

DOI: 10.1057/9781137542090.0007

This risk-neutral value is valid even if investors are risk averse (Brealey & Myers, 2003; Hull, 2009). By moving to an analytic framework that does not require estimated risk premia, the analyst escapes the very difficult task of estimating the risk of each component of a proposed investment at every point in time over the course of the valuation period. This is not to suggest that risk is unimportant in option valuation; rather, risk is captured in the value of the underlying asset at the time of the option's creation rather than in the option valuation process itself.

In summary, if the flexibility to take future action based on future conditions exists, a traditional benefit–cost analysis built on the discounted cashflow of net benefits may generate a misleading conclusion. When compared to the traditional discounted cash flow method, the real options approach offers several advantages. It permits a more sophisticated treatment of risk, avoids the need to estimate a series of risk premia applicable over the life of the R&D project, incorporates the managerial flexibility that is created by the execution of the R&D project, and reflects the likely evolution of the value of the underlying asset. As Myers succinctly put it: "DCF is no help at all for pure research and development. The value of R&D is almost all option value" (1984, p. 135).

Myers's comment notwithstanding, as shown in Chapter 4, real options analysis is not a panacea for all analytic difficulties. The method requires virtually all of the same inputs as the DCF method, plus some additional inputs that must be carefully developed.

3.2.2   Enhancement #2: treat groups of R&D projects as portfolios

The second potential enhancement to NRC's DT/DCF approach entails taking a portfolio perspective rather than looking at R&D projects in isolation. The Science of Science Policy Task Group concluded that metrics for judging the value of R&D portfolios are fully "relevant" to the challenges faced by Federal managers, but that there are "substantial gaps" in the maturity of the available methodologies and analytic inputs (ITG, p. 27). The Task Group did not, however, specifically identify those gaps, but the value of a strong portfolio management was evident in a survey of 205 firms active in product development:

> Those businesses that have gone to the trouble of installing a systematic, explicit portfolio management system—one with clear rules and procedures, that is consistently applied across all appropriate projects and *treats all projects*

as a portfolio, and which management buys into—are the clear winners. Their portfolios outperform the rest... *(emphasis added)*. (Cooper, Edgett, & Kleinschmidt, 2001, p. 371)

Abundant evidence suggests that most portfolio managers share a similar objective: to extract the maximum value—however defined—from their investment in the portfolio as a whole, rather than from any one investment in the portfolio (Evans, Hinds, & Hammock, 2009; Dias, 2006; Giebe, Grebe, & Wolfstetter, 2006; Cooper, Edgett, & Kleinschmidt, 2001). The central question here is therefore whether the key characteristics of an R&D portfolio are nothing more than the sum of the characteristics of the projects in it, or whether the construct of a "portfolio" connotes something more.

Point of departure: portfolios of financial investments

To understand portfolio effects, one instinctively starts with Markowitz's Nobel-prize winning work, in which he laid out a comprehensive theory of how to form a maximally efficient portfolio (i.e., a portfolio that, for a given level of risk, maximizes return, or alternatively, for a given level of return, minimizes risk) (1952). If a portfolio's most important attributes— its aggregate risk and its expected return—could be fully characterized based only on the attributes of the individual investments in the portfolio, then it would display both return additivity and risk additivity. In other words, the risk and return of the portfolio would be nothing more than the sum of the risk and return of the individual investment projects. If such a condition were to hold, there would be little value in studying "portfolios" as a construct distinct from the "projects" within it. As Markowitz convincingly demonstrated, however, risks are not additive, and as others have documented, at least when it comes to R&D, returns may not be additive either.

As regards risk additivity, Markowitz proved that the size of the portfolio and the relationships among the investments in it have a profound effect on the risk of the portfolio. When it comes to portfolio size, Markowitz's framework provides a mathematical structure for the advice that "one should not put all their eggs in the same basket." Markowitz demonstrates that the expected return, E, of a portfolio is

$$E = \sum_{i=1}^{N} X_i \mu_i$$

(1)

DOI: 10.1057/9781137542090.0007

and that the variance, V, of the portfolio's expected returns is

$$V = \sum_{i=1}^{N} \sum_{j=1}^{N} \sigma_{ij} X_i X_j$$

(2)

where N is the number of investments in the portfolio, X_i is the weight of investment i in the portfolio, μ_i is the return on investment i, and σ_{ij} is the covariance between the returns of investments i and j. The covariance term can be further decomposed:

$$\sigma_{ij} = \rho_{ij} \sigma_i \sigma_j$$

(3)

where ρ_{ij} is the correlation coefficient between the returns of investments i and j, and σ_i is the standard deviation of the returns of investment i. Markowitz himself does not equate portfolio variance with portfolio "risk." Bernstein, however, observes that "risk and variance have become synonymous" in common usage, although he questions "whether variance is the proper proxy for risk" (Bernstein, 1998, pp. 252, 257). For now, I accept the premise that risk can be characterized as the variance of expected returns (but return to this topic in Section 4.3.4).

One important implication of this framework is that increasing the number of projects in a portfolio, even if each project is identical with respect to risk and return, will lower the risk of the portfolio as a whole. Figure 3.3 demonstrates this effect in graphic form. I postulate an investment, the expected value of which is $100, with a standard deviation of $10. In the first case, the portfolio comprises a single investment. In a second, third, and fourth cases, the portfolio is split equally between, respectively, 5, 25, and 50 instances of the investment. To predict portfolio performance, I simulated four portfolios over 750,000 iterations.[7] The results of the projects are independent of each other, that is, the correlation coefficient, ρ, for each pair of projects is 0.0.

For all four portfolios, the return is constant at $100, the same as for the individual project. The standard deviation, however, drops from $10.00 for the single-project portfolio to $2.00 for the 25-project portfolio. In other words, increasing the size of a portfolio—even with additional projects of identical risk and return characteristics—reduces the variance of the expected returns while leaving the mean expected return unchanged. The practical implication of this phenomenon is a reduction in the uncertainty about the portfolio's performance. In the case of the single-project portfolio, the 95 percent confidence interval (CI)—the

DOI: 10.1057/9781137542090.0007

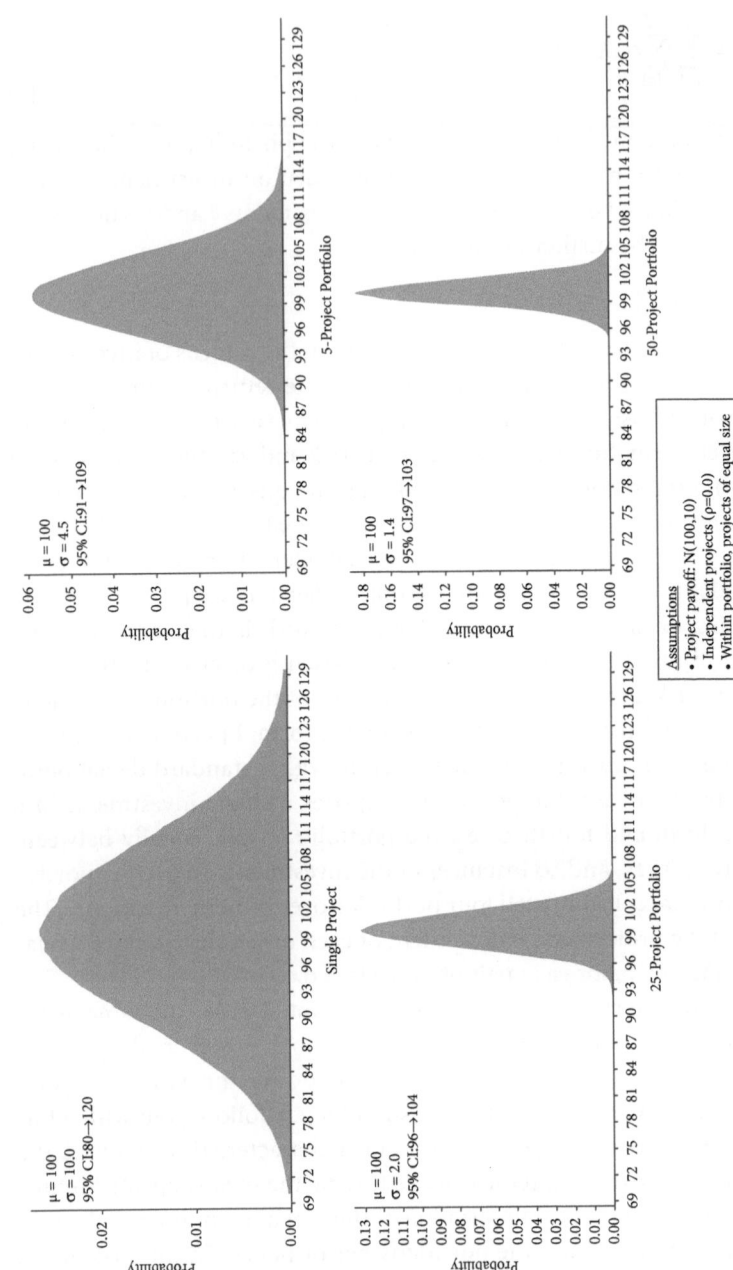

FIGURE 3-3 *Relationship between number of projects and portfolio variance*

DOI: 10.1057/9781137542090.0007

range within which the laws of probability suggest the ultimate outcome will lie—runs from $80 to $120. In the 25-project portfolio, the 95 percent CI is appreciably narrower, running from $96 to $104. This benefit to diversification accrues because of the offsetting effects of variations in project-level performance. Some projects will likely exhibit returns below the mean while others will like show returns above the mean; these variations tend to counterbalance one another, thereby reducing the variance of the portfolio's return.

Inspection of Equations (2) and (3) suggests that, in addition to the number of investments in a portfolio, another key driver of aggregate risk is the correlation among the performance of those investments. As the correlation coefficient, ρ, between any two investments in the portfolio decreases, the value of Equation (3) decreases, meaning that the associated covariance also drops. A lower covariance term in Equation (2), in turn, implies that the variance of the portfolio's returns will become smaller.

Figure 3.4 illustrates this effect in the case of a two-project portfolio. As before, the two projects are identical with respect to their expected value ($100) and standard deviation ($10). However, instead of assuming $\rho = 0.0$, I vary ρ from +1.0 to −1.0 in increments of 0.20, and simulate 50,000 iterations of each case. The results demonstrate that, all else equal, ρ has a significant effect on portfolio risk. In the limiting case,

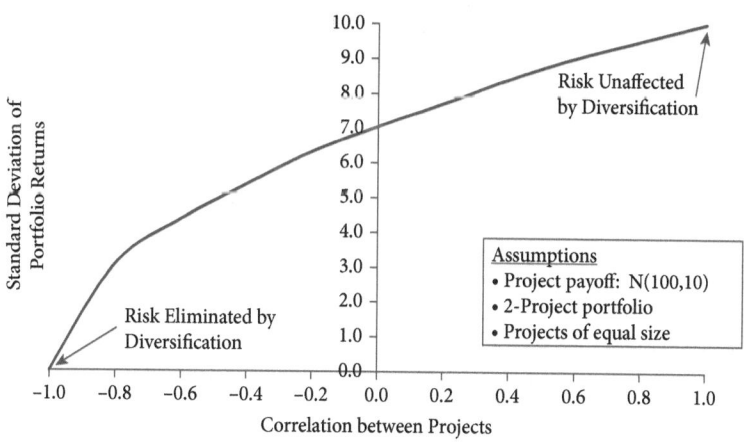

FIGURE 3.4   *Effect of project correlation on portfolio risk*
Source: Linquiti (2012b).

DOI: 10.1057/9781137542090.0007

when $\rho = -1.0$, the portfolio standard deviation is 0.0 and the portfolio is said to be perfectly hedged (i.e., performance can be predicted with no uncertainty). The intuition here is that regardless of the performance of the first investment, the second will perform in an equal but opposite fashion, thereby offsetting any deviation from the mean by the first.

As the correlation is increased from -1.0 toward $+1.0$, the impact of the hedging effect (i.e., risk reduction) on the portfolio standard deviation begins to diminish. Importantly, however, the hedging effect is observed in all cases except where $\rho = 1.0$. In that case, the returns of the two investments co-vary in an identical fashion. As a standard finance text puts it: "even if the covariance term is positive, the portfolio standard deviation *still* is less than the weighted average of the individual security standard deviations, unless the two securities are perfectly correlated" *(emphasis in original)* (Bodie, Kane, & Marcus, 2009, p. 199). In short, because portfolios fail to exhibit risk additivity, the construct of a "portfolio" is clearly more than the sum of the projects in it. This is true irrespective of whether the portfolio comprises financial investments or R&D projects.

Differences between porfolios of financial investments and of R&D projects

While Equation (2) and the analysis above demonstrate that the risks of specific investments within a financial portfolio cannot simply be added together to calculate portfolio risk, the same is not true for returns. In a financial portfolio, as is evident from Equation (1), the expected portfolio return does not depend on the existence of correlations among investments in the portfolio. As long as the portfolio comprises arms-length financial investments (e.g., stocks and bonds where the portfolio owner exerts no managerial control over the activities of the entities in which the investments have been made), expected returns are simply additive—there is no portfolio effect on returns. Even if the values of the financial investments in a portfolio are correlated, this correlation will affect only the risk of the portfolio, not its expected return.

In contrast, if an R&D portfolio is owned by a single entity (i.e., where there is not an arms-length relationship between the investor and the set of R&D investments, but instead where the portfolio owner can exert managerial control over the selection and direction of R&D activity), then unlike the financial portfolio, the expected return of the R&D portfolio may not simply be the sum of the expected return of all the R&D

DOI: 10.1057/9781137542090.0007

investments in it. Several scholars have investigated situations where this might occur. (Colvin & Marvelias, 2011; Wouters, Roorda, & Gal, 2011; Solak, Clarke, Johnson, & Barnes, 2010; Dias, 2006; Blau, Pekny, Varma, & Bunch, 2004; Triantis, Forthcoming; Chien, 2002; Childs & Triantis, 1999). An integrated reading of these sources suggests several ways in which the returns to an R&D portfolio may behave differently from a portfolio of financial assets such as stocks and bonds.

Interdependence of Commercial Returns: In some cases, the commercial value of an R&D project depends not only on its own results but also on the results of other, related projects. Such relationships may be constructive in the case of product complements, where the success of one product enhances the prospects of commercial success for another product, or may be destructive, as in the case of competing products where the success of one diminishes the success of the other. R&D of a new drug, accompanied by the development of a new delivery system, would be examples of the outcomes of complementary R&D. On the other hand, two new drugs that successfully treat the same disease would likely be substitutes in the marketplace, with one undermining the value of the other. Such interdependencies might be arrayed on a continuum that ranges from complete complementarity to mutual exclusivity. What's more, such interdependencies may not be symmetric. Consider, for example, the case of infratechnologies, defined here as industrial processes upon which multiple new products depend.[8] Failure in an R&D initiative targeted at an infratechnology virtually guarantees that R&D related to downstream innovations that depend on that infratechnology will also fail to yield valuable results (i.e., because the requisite infratechnology is absent), but the converse is not true. The failure of R&D for the downstream technology would not prevent success in the development of the infratechnology.

These sorts of potential commercial interdependencies can affect the expected return of the individual investments in an R&D portfolio differently than the individual investments in a financial portfolio. In the latter case, a decision to add to a portfolio an investment in a firm that competes with a firm already in the portfolio (e.g., to add stock in Coke to a portfolio that holds Pepsi stock) would not affect the expected return of the first stock (i.e., the return on the Pepsi stock does not depend on whether Coke is also in the portfolio). The same is not true for an R&D portfolio. If a pharmaceutical company invests in two R&D projects for drugs to fight the same disease (i.e., in competing products), the

DOI: 10.1057/9781137542090.0007

introduction of the second project into the portfolio would likely reduce the expected return of the first R&D project. The computation of the expected return on an R&D portfolio that includes either complementary or competitive projects is thus not the simple summation embedded in Equation (1). Instead, a more complex investigation of the expected returns of individual R&D investments, contingent on the nature of other investments selected for inclusion in the portfolio, is necessary.

Economies of Scope: In the typical portfolio of financial assets, the aggregate investment cost is simply the sum of the cost of acquiring each individual security. The same is not necessarily true in a portfolio of R&D investments, which may display economies of scope. If two (or more) R&D projects can share resources, in the form of facilities, personnel, or data, the aggregate cost of executing those projects is likely to be less than the sum of the cost of executing each project in isolation. And as costs drop, the return on investment goes up.

Endogeneity of Learning: The manager of a financial portfolio has only one (legal) way of learning about the returns of the investments in the portfolio—by observing changing market prices over time. Such learning takes place exogenously (i.e., from outside the portfolio), but the same is not true for an R&D portfolio where knowledge learned in one R&D project can affect management decisions about that or other R&D projects. Early technical success in one project, for example, may prompt management to increase the investment in it, thereby speeding project execution, while simultaneously reducing or eliminating spending on a parallel project. Management may also choose to sequence projects in series, knowing that the results of one early R&D project may inform (and improve the return on) multiple subsequent projects. The fact that endogenous learning can improve the returns from a portfolio and depends on the specific composition of the portfolio is further evidence that, when it comes to R&D investments, portfolio return is not the simple sum of project returns.

Given that risk additivity never exists within any portfolio (except for one comprised of perfectly correlated assets) and that return additivity is often absent from R&D portfolios, it is important to consider the relationship between risk and return. One last concept from Markowitz's work is relevant here—the concept of portfolio efficiency. Consider, as shown in Figure 3.5, a large number of portfolios that could be formed by various combinations of investment securities. In this framework, each portfolio is described in terms of its expected

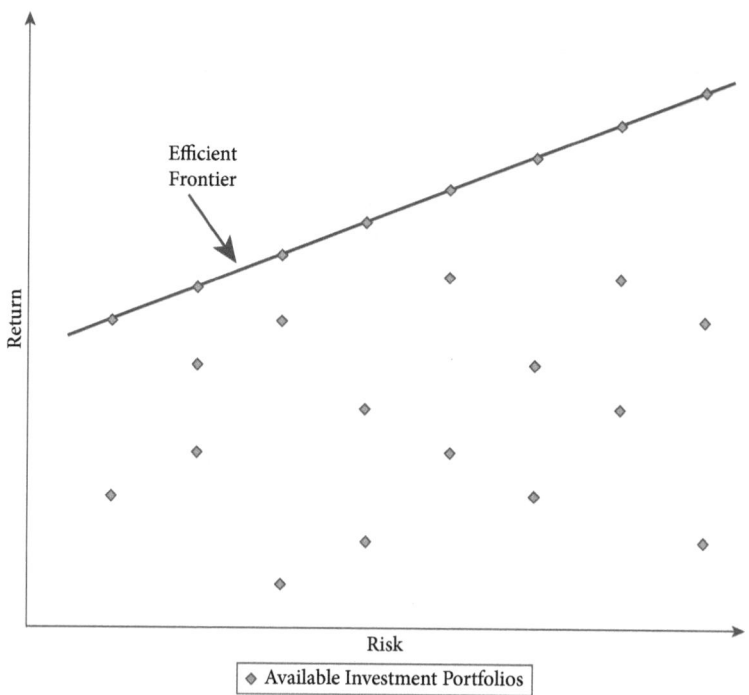

FIGURE 3.5   *Trade-off of risk & return*
Source: Linquiti (2012b).

return and its variance. Markowitz demonstrates that it is not possible to simultaneously maximize returns while minimizing variance, as these two objectives are in tension with one another. What's more, without knowing the investor's preferences for risk, the analyst cannot identify a single optimal investment portfolio but instead, can only identify the set of efficient portfolios, defined as those portfolios that, for a given level of risk, maximize the return, or conversely, for a given return, minimize risk. Such portfolios form an efficient frontier. The frontier is typically assumed to be upwardly sloped owing to investors' aggregate risk aversion; to entice investors to take more risk, a riskier investment must offer a higher potential return. Irrespective of risk aversion, however, a rational investor would only select portfolios on the frontier; any other choice would be suboptimal because there would be another investment with the same risk, but higher returns, or the same return, but lower risk.

DOI: 10.1057/9781137542090.0007

Markowitz provides a framework for forming efficient portfolios by blending various classes of risky assets with long or short positions in risk-free bonds. The specific details of Markowitz's framework are not germane to the analysis here. What is most important is that policymakers building R&D portfolios are well advised to consider both risk and return, and to always prefer efficient portfolios (Markowitz, 1952; Linquiti, 2012a). The desirability of purposively trading off risk and return is yet one more reason why a portfolio ought to be viewed as more than a simple aggregation of the projects in it.

3.2.3   Enhancement #3: value R&D portfolios as portfolios of real options

While portfolio analysis—applied to stocks and bonds—has a decades-long history of study and practice, the same cannot be said for the analysis of portfolios of real options. Anand, Oriani, & Vassolo note: "the literature is still in its initial stages regarding portfolios of strategic investments.... [T]he issue of potential interactions among real options has so far scarcely been investigated" (2007, pp. 276, 279). Other investigators have reached similar conclusions (Eliat, Golany, & Shtub, 2006; Childs, Ott, & Triantis, 1998; Sick & Gamba, 2005).

Intuitively speaking, the fact that a real option bounds the downside loss means that the effect of diversification is fundamentally changed. In a portfolio of assets that all have similar upside and downside risk, above-average payoffs among some assets will tend to be offset by below-average payoffs of other assets (thereby lowering portfolio risk). The same is not true for options portfolios, where above- and below-average payoffs are not symmetrically distributed. van Bekkum, Pennings, & Smit confirm this intuition:

> [T]he conditionality of investment decisions in R&D has a critical impact on portfolio risk, and implies that traditional diversification strategies should be reevaluated when a portfolio is constructed.... Although the risk of a portfolio always depends on the correlation between projects, a portfolio of conditional R&D projects with real option characteristics has a fundamentally different risk than a portfolio of unconditional risk. (2009, p. 1150)

To demonstrate their point, van Bekkum et al., use a Monte Carlo model to illustrate portfolio outcomes. Borrowing their methodology, and applying it to the example that has been developed in this chapter, Figure 3.6 shows the effect of conditionality on portfolio risk. As in Figure 3.4, I

DOI: 10.1057/9781137542090.0007

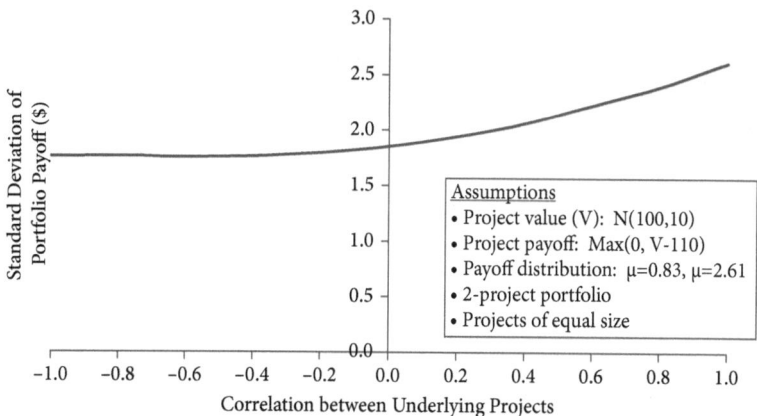

FIGURE 3.6   *Effect on portfolio risk of correlation between projects with conditional payoffs*

simulate (for 50,000 iterations) a portfolio of two identical projects with μ = \$100 and σ = \$10; in this case, however, the projects yield a payoff only if their underlying value (V) exceeds \$110. The payoff is V-110.

The differences between Figures 3.4 and 3.6 are stark. The vertical scales are different, because the unconditional project has σ = \$10, while the payoff from the conditional project has σ = \$2.60. That notwithstanding, the effect of combing two investments, the returns from which are negatively correlated, has a much smaller impact on the risk of an option portfolio than it does on a portfolio of unconditional investments. The standard deviation of the portfolio drops from about \$2.60 for two perfectly correlated investments to about \$1.90 for two independent projects (i.e., ρ = 0.0). As ρ becomes progressively more negative, however, the standard deviation plateaus at an asymptote of about \$1.80. In other words, in this simple case of two identical conditional assets, it is not possible to fully hedge the portfolio against risk (in contrast to the unconditional project case, where in the limit of ρ = –1.0, a perfect hedge is created). Accordingly, van Bekkum et al., add a cautionary note: "conventional diversification arguments do not hold since conditionality causes the payoffs to become nonlinear [and] reliance on traditional diversification strategies can be misleading" (2009, p. 1150).

To recap, the certain absence of risk additivity in all investment portfolios, the frequent absence of return additivity in R&D portfolios, the

DOI: 10.1057/9781137542090.0007

value of purposively trading off risk and return, and the complex inter-action of investments with conditional payoffs are all persuasive reasons to analyze and value not only individual R&D projects, but also the R&D portfolios they comprise.

3.3   Approaches to portfolio formation

Practitioners and scholars have attempted to tackle some of these unique features of R&D investments and incorporate them into recommenda-tions about how to form and value R&D portfolios. Below is a quick review, first of commercial practices and then of public sector practices, for portfolio formation.

3.3.1   Commercial practices for portfolio formation

In the private sector, techniques for combining individual R&D projects into a portfolio typically apply one of four approaches. First, they may take a qualitative perspective on multiple objectives to inform portfolio formation. Second, they may focus on multiple objectives, but apply quantitative methods. Third, they may address a single objective—usually profit or value maximization—and apply quantitative methods to identify the optimal portfolio as of the date of portfolio formation. Finally, some investigators have framed R&D portfolio management as a dynamic problem and explicitly considered how active management of a portfolio can increase its value.[9]

The bucket method

The first approach to R&D portfolio management is a qualitative one that might be termed the "bucket method" in which candidate R&D projects are sorted into categories (i.e., buckets) based on a taxonomy of charac-teristics, such as process versus product change, degree of technical risk, magnitude of innovation (e.g., radical vs. incremental), relationship to existing products and markets, and so forth. Managers then endeavor to align the buckets with overall organizational strategies and allocate resources accordingly. A mature business, seeking to defend an exist-ing market position, may emphasize incremental process and product improvements in its current operations. By contrast, a start-up seeking to distinguish itself in a crowded marketplace might focus almost exclu-sively on radical, potentially disruptive, new technologies. Successful execution of the "bucket method" requires definitions of buckets that can discriminate between R&D projects that align well with the firm's business strategy and those that don't.

MacMillan and McGrath provide an example of a bucket method that explicitly incorporates real options thinking (2004). They argue that firms must select among five basic types of R&D projects that are distinguished by the extent of both market and technical risk that they carry. Two types of R&D carry such modest levels of risk that, according to MacMillan and McGrath, option analysis is not particularly useful. "Enhancement launches" and "platform launches" both carry only low or medium risk and are closely related to a firm's existing lines of businesses. As such, firms are advised to simply design and launch these new initiatives as one integrated decision. With low or medium risk, and a relatively short time to market, there is little opportunity for "learning" and hence the key driver of an option's value is not present.

In contrast, the other three types of R&D investment carry significant risk and are best conceptualized within an options paradigm. MacMillan and McGrath characterize projects with high technical and market risk as "Stepping-Stone Options," projects with high technical risk and low or medium market risk as "Positioning Options," and projects with high market risk and low or medium technical risk as "Scouting Options." To manage a Positioning Option, the firm needs to resolve technical uncertainty and will need to make several small investments in competing technologies to understand which may prove most fruitful for further development. On the other hand, the emphasis with a Scouting Option is to resolve market, rather than technical, uncertainty and thus key information is likely to lie outside the firm's laboratories and development facilities. Learning about customer needs and competitor behavior is the key to reducing the uncertainty associated with a Scouting Option. Finally, Stepping-Stone Options require the simultaneous reduction of technical and market uncertainties. The firm should thus make R&D investments that reduce both types of uncertainty in tandem.

Schilling (2010) provides a different configuration for the implementation of the "bucket method." In her framework, R&D projects are characterized on the basis of two dimensions: the magnitude of the potential product change and the magnitude of the potential process change. Combining these two dimensions, R&D initiatives are sorted from less radical to more radical: (1) derivative projects, (2) platform projects,

(3) breakthrough projects, and (4) advanced R&D projects. Depending on its overall strategy, Schilling suggests that a firm then decide how much of its R&D budget to allocate to each bucket. A firm with short-term cash flow requirements might devote most its resources to derivative projects while a firm aiming for high growth might focus exclusively on breakthrough and platform projects.

A shortcoming of the bucket method is that, taken by itself, it does not help an investor compute the optimal amount to spend on R&D. Instead, the overall budget constraint (i.e., the funds available for R&D) is specified exogenously. Accordingly, the bucket method, while likely improving the cost-effectiveness of R&D spending, should be considered a boundedly rational approach to decision-making. Applying standard microeconomic thinking, a firm would want to fund R&D up to the point at which the marginal benefit of the investment is equal to its marginal cost (Evans, Hinds, & Hammock, 2009; Childs, Ott, & Triantis, 1998; Davis & Owens, 2003). The bucket approach does not help in making this calculation.

In addition, while the bucket method does not intrinsically prevent the analyst from considering potential correlations among the technical and commercial performance of different R&D projects—either within the same bucket or across buckets—the literature on the bucket approach is generally silent on the matter of project relationships.

Quantitative, multi-objective, methods

When confronted with a choice among several investment options that vary with respect to multiple decision-relevant attributes, a decision-maker must apply some method for assimilating the available information to identify the preferred investments. In the context of R&D, candidate projects may have several relevant attributes, such as market and technical risk, commercial payoff, fit with corporate strategies, and so forth. One standard approach in decision sciences is to develop a scoring system (perhaps from 1 to 10) for each attribute and then score each alternative investment accordingly. A second step is to weight each attribute in terms of relative importance (such that the weights sum to 1.0). Finally, the scores and weights are combined to provide a weighted score with the interpretation that the higher the score of an individual option or investment, the closer it matches the decision-maker's preferences (Ragsdale, 2008). A linear programming model can then be used to select investments that maximize

DOI: 10.1057/9781137542090.0007

the aggregate scores across a portfolio subject to constraints, such as budget limits or the availability of human or physical resources. While examples of this approach in the valuation of R&D projects do exist (Peerenboom, Buehring, & Joseph, 1989; Golabi, Kirkwood, & Sicherman, 1981), it does not appear to be in widespread use for R&D portfolio formation.

One concern with such approaches is the subjective nature of the scoring systems and of the weights assigned to the relevant attributes, particularly when decision-makers are confronted with a large set of candidate investments that must be scored simultaneously in a consistent fashion (Linton, Walsh, & Morabito, 2002). A related approach, Data Envelopment Analysis (DEA), is thus often suggested as an alternative (Schilling, 2010). Using DEA techniques, the decision-maker is relieved of the need to specify project scores and attribute weights. Instead, the DEA method leaves project characteristics in their natural units (rather than converting them to a 1 to 10 scale) and uses linear programming techniques to identify the set of attribute weights that maximize the apparent efficiency of each candidate investment. Projects are then measured on the basis of their distance from the most-efficient project. With a single metric for each project, projects can quickly be screened in or out of a portfolio selection process, with projects exhibiting mid-range scores being reserved for more detailed review outside the DEA framework (Linton, Morabito, & Yeomans, 2007).

While it does help the portfolio manager deal with a diverse set of qualitative and quantitative attributes, the DEA method does not readily accommodate the simulation of relationships among the candidate projects, a significant shortcoming. In addition, in the examples reviewed herein, the DEA approach is not integrated with a real options perspective that captures managerial flexibility in the face of evolving uncertainty, although, in principle, there is no reason why option value could not be used as a project attribute in a DEA analysis.

Static choice methods

Another, strictly quantitative, approach to forming R&D portfolios grows out of the field of decision sciences and dates to the development of linear programming methods during World War Two (Wagner, 1975). Such studies focus on maximizing a single objective—usually profit or market value—subject to constraints related to budgets, allocation across buckets, and sometimes, risk. In this approach, the optimal portfolio(s)

DOI: 10.1057/9781137542090.0007

are identified as the outset of the planning process. While they invariably consider a multiyear planning horizon, such models are typically not a real-time management tool, although they can be updated frequently to reflect new information and to re-assess an existing portfolio. Owing to the nonlinearity of the portfolio variance term in Equation (2), portfolio optimization analysis is typically done using nonlinear programming methods (Ragsdale, 2008). Such methods can take account of correlations among investments and are used to identify, for a given target portfolio return, the set of investments that will produce the lowest variance in the portfolio return. By repeating the model analysis for different target returns, the analyst can construct an efficient frontier of portfolios of the type defined by Markowitz.

If the formation and operation of a portfolio will be driven by multiple nonlinear relationships, discontinuous functions, probability distributions that do not match the "standard" library of distributions, and/or complex decision rules, then stochastic simulation in a Monte Carlo model is often the approach preferred by analysts (van Bekkum, Pennings, & Smit, 2009; Bodner & Rouse, 2007; Sick & Gamba, 2005; Triantis, 2003).

Regardless of the analytic model employed, the objective is typically the same: to identify one or more R&D portfolios that lie on the efficient frontier. The following three examples demonstrate how this approach is typically applied.

The first example is taken from the aerospace industry where a nonlinear optimization model was used to select projects for development (Dickinson, Thornton, & Graves, 2001). This case is noteworthy because, in addition to project cost, revenue, and risk, dependencies among projects were explicitly modeled; if two projects were deemed complements in the marketplace, then the commercial value of one would be reduced if the other was not selected for the portfolio (and vice versa). The analysts also allowed for asymmetries in the project relationships, so that the reduction in commercial value did not have to be the same in the case of the two complements. Another constraint in the optimization model was that selected projects had to achieve certain strategic corporate objectives. The model builders worked closely with corporate product development engineers to specify model inputs and analyze "what-if" scenarios under which the inputs were changed and consequences for the portfolio as a whole examined. The model, however, applied a standard discounted cash flow valuation, rather than a real options valuation.

DOI: 10.1057/9781137542090.0007

An example of a stochastic model application is the use of real option techniques to optimize a series of go/no-go decisions being made under a budget constraint across a pipeline of pharmaceutical R&D projects (Rogers, Gupta, & Maranas, 2002). The study demonstrates a typical finding in the real options literature—that seemingly unattractive investments in low probability, high payoff, projects with flexibility may in fact contain significant value. Novel about this example is that the investigators tackle a reasonably complex demonstration case, with a collection of 20-candidate R&D project spread across six stages of product development, yielding almost 13,000 continuous variables and another 900 binary variables in the optimization model. Real options are valued using a risk-neutral decision-tree approach built around both market and technical risk, although the authors do not model correlations among the projects in the pipeline.

A final example of the static choice approach comes from a simulation of R&D projects in the forest products industry over a six-stage product development lifecycle (Bodner & Rouse, 2007). Here, the simulation is not stochastic but instead relies on a discrete-event simulation that implements a set of user-supplied decision rules, including a budget constraint and management preferences for work in certain product lines. R&D projects are modeled as real options and valued using the Black–Scholes–Merton model.[10] The commercial value of all projects is assumed to be subject to the same variability over time, and correlations are not modeled between projects. Nonetheless, the investigators demonstrate that decision rules built on options thinking suggest higher portfolio returns than rules built around maximizing the net present value of expected cash flows.

Dynamic management methods

Another strand of research in the field of optimizing R&D portfolios has focused on the dynamic interactions among R&D projects of the type described in Section 3.2.2 (e.g., endogeneity of learning, economies of scale and scope in R&D project specification, and complementarity and substitutability in the marketplace) and the opportunities for ongoing strategic management of the portfolio in light of these interactions. By way of illustration, I describe below two prominent examples from this literature.

In two related articles, Childs and his co-authors investigate the optimal sequencing of the formation and exercise of real options in R&D and related capital investment policies (Childs & Triantis, 1999; Childs, Ott, & Triantis, 1998). In the first article, the investigators postulate a firm with two development projects, only one of which can be implemented. They demonstrate that, with endogenous learning, the optimal development strategy depends on the relationship between the two projects. If they are highly correlated, sequential development may be superior to parallel development, especially if capital requirements are high and development times short, because the results of the first project can be used to inform the execution of the second.

The second article is more expansive, considering a range of additional factors (such as competitor behavior, capital rationing, and correlations in project payoffs) and more explicitly emphasizing the dynamic management of a collection of real options. Using a decision-tree-based approach to real option valuation, the authors investigate multiple strategies for a two-project portfolio: parallel versus sequential development, hybrid strategies where managers dynamically switch between parallel and sequential development based on ongoing results, and strategies in which projects are accelerated or slowed based on technical progress and evolving market conditions. Childs and his colleagues demonstrate that not only do the relationships among projects matter when it comes to portfolio value, the dynamic management methods applied to the portfolio are also an important driver of portfolio value.

Smith and Thompson investigate the trade-off between diversification and concentration of real options within a portfolio, with a focus on the petroleum exploration industry (2009). They observe that because of the possibility of endogenous learning (e.g., about the prospects of finding oil in a particular area), it may be desirable to develop concentrated options, the values of which are closely dependent on each other. Their empirical investigation of a series of lease sales on the Outer Continental Shelf in 1973 and 1974 suggest that firms often display such preferences. Such an observation is at odds with the typical advice that adding assets to a portfolio that are negatively correlated with other portfolio assets is a mechanism for reducing risk. Smith and Thompson's work, however, suggests that if the real options have a sequential, or serial, nature, where the learning from project can be used to inform the next project in the sequence, then risks may be reduced. Conversely, if all projects must operate in parallel, the opportunity for endogenous learning would be substantially reduced, thereby limiting the value of concentration.

DOI: 10.1057/9781137542090.0007

3.3.2 Public sector practices for portfolio formation

In contrast to the typical private sector practice of applying a strategic, portfolio-wide, perspective to R&D management, government policymakers tend to focus on the merits of individual projects without consideration of portfolio effects (Litvinchev, Lopez, Alvarez, & Fernandez, 2010; Giebe, Grebe, & Wolfstetter, 2006; Bozeman & Rogers, 2001). A senior Congressional aide remarked that the US Federal R&D system is generally controlled from the "bottom up, not the top down" (Sokolov, 2012).

The literature does include some examples of the application of a portfolio-wide, rather than a project-oriented perspective, to Federal R&D programs, although they appear to be the exception rather than the rule. Both the Army and the Navy, for example, have used a tool known as "Port-Man" to characterize and select R&D projects. Applying linear programming techniques, the tool values projects based on how well each might close gaps between various military mission requirements and existing military capabilities. The model constrains outcomes (i.e., R&D project selections) to ensure that multiple mission objectives are addressed and that budgets are met. The risk of technical failure is explicitly incorporated in later versions of the tool, but potential correlations among the technical results of the projects or of their military value are not considered (Silberglitt et al., 2004; Chow, Silberglitt, & Hiromoto, 2009; Chow, Silberglitt, Hiromoto, Reilly, & Panis, 2011).

The US Army Corps of Engineers' Engineer Research and Development Center (ERDC) manages its R&D program with a systematic "bucket" approach for selecting and managing R&D projects. Each candidate R&D project is evaluated on the basis of how well it aligns with the Corps' overall military or civil works mission objectives, whether it augments an "aging" or "emerging" technology, whether it serves a business area in decline or of high national priority, and the degree of technical risk (from "not likely" to succeed to "no brainer"). Projects are then classified into one of four "buckets" based on whether they score "high" or "low" on two dimensions: mission relevance and momentum. ERDC endeavors to cull projects with low momentum and mission relevance from the portfolio, and increase investments in those with high momentum and mission relevance (Holland, October 29, 2009).

An extremely broad perspective on the factors that affect the design and assessment of public research portfolios is provided by Jordan

DOI: 10.1057/9781137542090.0007

(2011). In her framework, relevant actors, institutions, and interactions are characterized at three different levels: the macro, the meso, and the micro. At the macro-level, key factors include national policies, the state of the macroeconomy, the availability of risk-tolerant capital, and existing modes of coordination among high-level actors. At the meso-level, technological and institutional considerations are paramount and include research facilities and infrastructure, individual researchers and the networks among them, information infrastructure, supply chain capabilities needed to bring new technologies to market, government infrastructure, and end-user demand. At the micro-level, R&D projects are sorted into one of four "buckets" depending on their scope (broad vs. narrow) and distance from status quo technologies (radical vs. incremental): (1) narrow/radical "seize the day/be new," (2) broad/ radical "force for the future/be first," (3) broad/incremental "parts for the whole/ be better," and (4) narrow/incremental "it grows as it goes/be sustainable" (pp. 13–14). While quite broad—indeed broader than my focus here on the valuation of specific R&D portfolios—Jordan's framework is a useful reminder that public sector R&D exists with a complex and dynamic system. As demonstrated in the Chapter 2 review of a logic model for public sector R&D programs, a policymaker interested in effective R&D portfolio management needs to be mindful of and capable of addressing the multiple factors that affect the performance of this system.

Several scholars have also developed more focused methods for forming and/or valuing R&D portfolios in the public sector and then applied those methods to a particular case study. For example, Davis and Owens use real option techniques to investigate investments in renewable electric R&D by DOE (Davis & Owens, 2003). Vonortas and Hertzfeld describe a real option method and illustrate it with an application to NASA's High Speed Research Program (1998). A multi-attribute utility function was used to rank and select a group of environmental projects supported by a single 1982 funding tranche within DOE's synthetic fuels program (Peerenboom, Buehring, & Joseph, 1989). Golabi, Kirkwood, and Sicherman describe a similar initiative in which 77 R&D project proposals submitted to DOE's solar program were characterized in terms of 22 attributes and then, using an integer linear programming model, 17 projects were selected on the basis of a portfolio-wide multi-attribute utility function (1981).

DOI: 10.1057/9781137542090.0007

Notes

1   Much of the material in this chapter is based on my dissertation (Linquiti, 2012b).

2   As explained in Chapter 2, my focus is limited to *applied* research and development, meaning R&D initiatives the purpose of which is to generate new technologies that can be commercialized in the marketplace. Basic research, intended to generate new knowledge without regard to potential commercial application, poses additional analytic challenges when it comes to valuation. Such challenges are not addressed in this analysis.

3   In all cases, NRC assumed a counterfactual in which the private sector eventually would have developed the technology without government support. The stream of benefits is thus the incremental energy cost savings that result from accelerating deployment of the technology.

4   The 2005 NRC report suggests use of a three percent rate based on guidance issued by the Office of Management and Budget (OMB), but this suggestion is not followed consistently in the 2007 NRC report.

5   Using standard DCF methods to value an investment where risk, and hence the discount rate, is constantly changing can be analytically intractable (Hull, 2009, p. 243; Brealey & Myers, 2003, p. 591).

6   One exception might be a publicly traded start-up venture working to develop and commercialize a single technology.

7   The same results could have been obtained by applying Equations (1), (2), and (3), but as shown in the discussion to follow, use of the Monte Carlo approach facilitates a more sophisticated investigation of the resulting distributions.

8   See Tassey (2005) for a detailed description of the role played by infratechnologies in the process of technical innovation.

9   Brosch reminds us that designing the optimal portfolio "supposing optimal future exercise" of the real options in it and actually executing that portfolio by "exercising existing real options optimally" are two distinctly different tasks (2008, p. 15). My focus is here is on the former task, rather than the latter.

10 The Black–Scholes–Merton model is the foundation of option value techniques in the financial industry. It won its two of its developers the Nobel Prize. (Fischer Black died prior to the award of the Prize).

DOI: 10.1057/9781137542090.0007

4

Public Sector R&D Valuation: A Practical Example

Abstract: *Chapter 4 demonstrates the practical feasibility of prospectively using a real options approach to value a portfolio of government R&D projects characterized by significant technical, commercial, and other interdependencies and uncertainties. The studied R&D projects focused on energy consumption in the chemicals industry and were sponsored by the Department of Energy between 2002 and 2004. Results from this practical example confirm that switching from a discounted cash flow method to a real options method for valuing R&D investments makes a material difference at both the project and portfolio levels. In addition, the risks that are taken with taxpayer money—usually obscured by traditional methods—become plainly visible in the approach that is demonstrated.*

Keywords: DOE energy efficiency programs; portfolios of real R&D options; R&D project correlations; R&D project ranking; R&D risks

Linquiti, Peter D. *The Public Sector R&D Enterprise: A New Approach to Portfolio Valuation*. New York: Palgrave Macmillan, 2015. DOI: 10.1057/9781137542090.0008.

DOI: 10.1057/9781137542090.0008

With the NRC methodology as a point of departure, and given the preceding review of relevant theory and literature, I turn in this chapter to the practical task of prospectively estimating the monetary value of a set of government R&D projects.[1] I start with four research questions that guide the analysis and then describe the data and methodology that are used to answer those questions. Finally, I present and discuss the results of the analysis.[2]

4.1   Research questions

This chapter aims to answer several research questions by working through the calculations associated with different valuation methods for the same collection of "real-world" R&D projects.

▸ How do the results of a real option approach differ from those of a more traditional DT/DCF analysis? What are the policy implications of any such differences when it comes to selection and valuation of government R&D projects?

▸ How does moving from a project-level perspective to a portfolio-wide perspective affect the prospective valuation of R&D investments? What additional insights about potential risks and returns can be gleaned with a portfolio approach?

▸ How does the predicted performance of a portfolio of R&D projects (when viewed as real options) change when the relationships among the likely technical and commercial performance of those projects are characterized and modeled? Which of these relationships are amenable to control by public sector R&D managers, and which are not?

▸ What methodological issues are likely to arise with the application of real option techniques and the analysis of project interdependencies in the typical public sector R&D program?

With answers to such questions in hand, policymakers should be able to better judge the potential return on, and likely risks of, public investment in R&D and, in turn, make more informed choices about portfolio formation.

4.2   Data

This study re-analyzes the 22 R&D projects described previously in Section 3.1. These projects were sponsored by DOE's Chemical Industrial Technologies Program between 2002 and 2004 in an attempt to develop

DOI: 10.1057/9781137542090.0008

new technologies to reduce energy use in the chemicals industry, one of the largest energy-consuming sectors in the US economy. Table 4.1 provides a roster of these projects classified according to a technology-based taxonomy developed by DOE.

Each of these projects was aimed at fostering the commercialization of a particular new energy-efficient technology in the chemicals industry. As such, these projects did not constitute "basic" research. Some of the projects focused more on solving specific technical engineering challenges while others focused on developing readily deployable technologies; accordingly, this portfolio comprises a mix of "applied research" and "development." In all cases, however, NRC's technical experts were able to trace a direct line from the project to potential commercialization of a new technology and, in turn, to valuable reductions in energy use. Accordingly, further parsing the projects into the categories of "applied research" or "development" would not affect my proposed methodology, or the results presented.

4.2.1   R&D project data

Each of the R&D projects was characterized on the basis of its risks and potential payoffs, its cost and timing, and its relationship to other projects in the portfolio.

Project risks and payoffs: NRC data

The NRC convened a panel of industry experts to review each R&D project (NRC, 2007). After two rounds of deliberation, the panel estimated the probabilities of both market and technical success for each project. The estimated probabilities of technical success for the 22 projects ranged from 5 to 50 percent, with an average of 19 percent. The panel acknowledged that this binary approach (i.e., that a project is fully successful or fails completely) was an oversimplification of a reality that could include degrees of partial success.[3] Given available data, I was unable to develop a more nuanced characterization of the techical risk associated with these projects.

The panel also estimated the likely market success of each technology (conditional on technical success). Estimated probabilities of market success ranged from 10 to 68 percent, with an average of 31 percent. The panel assumed that if the new technology was successful in the market, then it would be deployed simultaneously across the industry. While the analysis presented later in this chapter maintain NRC's assumption about the instantaneous diffusion of the new technology, it does not take NRC's estimated probability of market success as an input. Instead,

DOI: 10.1057/9781137542090.0008

TABLE 4.1 *Taxonomy of R&D projects*

Focus area	Sub-focus area	Project #	Project Name
Reactions	Oxidation reactions	1	Using ionic liquids in selective hydrocarbon conversion processes
		2	Millisecond oxidation of alkanes
		3	From natural gas to ethylene via methane homologation and ethane oxidative dehydrogenation
		4	Development of high selective oxidation catalysts by atomic layer deposition
	Micro reactions	5	Micro-channel reactor system for hydrogen peroxide production
		6	Micro-channel reactor system for catalytic hydrogenation
		8	Process intensification through multifunctional reactor engineering
	New process chemistry/ Synthesis	9	Purification process for purified terephthalic acid (PTA) production
		10	Tackifier dispersions for pressure-sensitive adhesives
		11	Low cost chemical feedstocks using an improved and energy-efficient natural gas liquid removal process
		12	New sustainable chemistries of low VOC coatings
	Bio-catalysis	13	Production and separation of fermentation-derived acetic acid
Separations	Distillation & hybrid	14	Heat integrated distillation through use of micro-channel technology
	Membranes	16	Advanced membrane technology for hydrocarbon separations
Enabling technologies	Materials, computations, sensors, controls	18	Chemical industry corrosion management
		19	Enhanced productivity of chemical processes using dense fluidized beds
	Industrial energy systems	20	Paraxylene production with waste heat powered ammonia absorption refrigeration
		21	Waste heat powered ammonia absorption refrigeration unit for LPG recovery
		22	Rotary burner technology demonstration

Projects #7, #15, and #17 were removed from the analysis, for reasons described in the text.

Source: Fri (November 2009).

DOI: 10.1057/9781137542090.0008

the market success for a particular technology is predicted within the modeling framework based on the probabilistic simulation of energy cost savings and the cost of technology deployment.

In addition, working with DOE consultants, the panel also estimated the likely national energy savings by year that would result from implementation of each technology. These savings were expressed in natural physical units (e.g., million BTU) and represent the incremental difference between energy used with and without the technology. In all cases, the panel believed that the private-sector, acting without DOE's support, would eventually develop and deploy the technology, but would do so at a later date than if DOE had supported it. The panel assumed the same probabilities of technical and market success would apply in this counterfactual situation.[4] The benefits of the DOE program thus originate in its acceleration of the development and deployment of new technologies. For most projects, the acceleration is between 3 and 5 years, with one at 10 years and another at 20 years. The stream of energy savings is truncated in the year 2030; benefits after that point were not considered by NRC.

To monetize the energy savings of each project, the panel used a forecast of energy prices developed by the Energy Information Administration (EIA). The panel applied a heuristic rule of thumb that the cost of deploying the new technology would be three times the annual energy savings. With this assumption, and the panel's energy cost forecast, I was able to compute the deployment cost for each technology (i.e., the option exercise price). Costs are expressed as 2003 dollars.

I assumed that the date for deciding whether to deploy the technology would be the year prior to the NRC-estimated start of the energy cost savings (implicitly allowing one year for installation of the new technology). This date is the point when the decision to exercise the real option would be made (i.e., whether to deploy the new technology).

I dropped two projects from further analysis because NRC estimated negative cost savings for each (i.e., that installation of the new technology would actually increase costs) and thus the heuristic method of computing deployment cost was infeasible.[5] Much of the data used in this analysis were not presented by NRC in its report, but were provided directly to me by Robert Fri, the chair of the NRC panel (Fri, 2009).

One component of the NRC study did tangentially address portfolio effects. The study authors analyzed the 22 R&D projects with a Monte Carlo model that simulated portfolio results over 1,000 iterations. Other than reporting an expected value of $534 million (2003$, discounted

DOI: 10.1057/9781137542090.0008

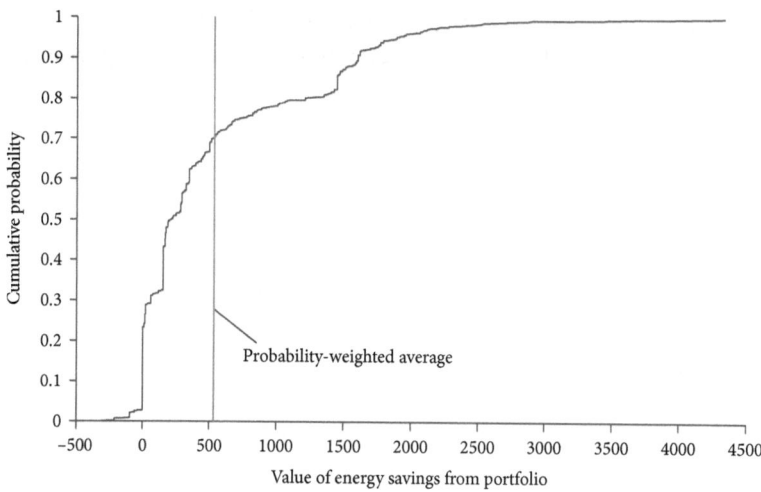

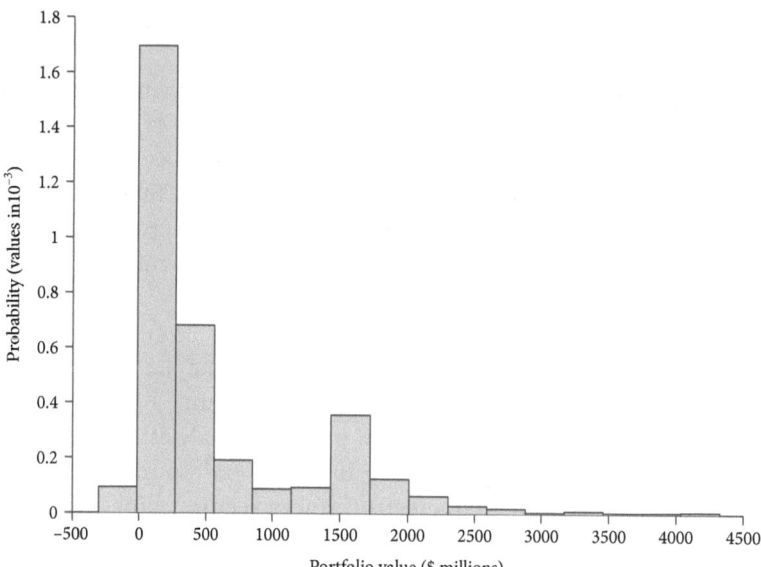

FIGURE 4.1   *"Uncertainty surrounding estimates of program benefits" as presented by National Research Council*

Source: NRC (2007, p. 213). Reprinted with permission.

DOI: 10.1057/9781137542090.0008

at 3%) and displaying the graphics shown in Figure 4.1, NRC did not further characterize the expected performance of this R&D portfolio.

Project cost and timing: DOE data

Administrative data were obtained from DOE (Ozokweleu, 2010). For each project, I was given the start and end date, along with the government and private-sector costs associated with executing the project. Some projects had NRC-predicted energy savings prior to the DOE-reported completion date for the R&D project. This anomaly arose because the NRC panel did not have details on the schedule for each project. In such cases, I deferred to the panel's judgment about the timing of benefits and reset the project end date to be one year prior to the start of savings (to allow one year to implement the newly developed technologies). R&D costs were assumed to be incurred uniformly over the duration of the project and were converted to 2003 dollars using a GDP deflator (US Office of Management and Budget, 2011). I assumed that projects were being valued in 2002 and all R&D costs are net present values as of 2002.

One project was excluded because it ended in 2003 which was inconsistent with a prospective analysis of it performed as of 2005 (the date of NRC's analysis). Because two projects had already been excluded, the result was 19 R&D projects in the data set for analysis. Table 4.2 summarizes the data, with the modifications described above, provided by DOE. These 19 projects entail R&D spending of about $67.54 million.

Project relationships: ICFI data

NRC did not consider relationships among the R&D projects in the Chemical Industrial Technologies Program. Because of the importance of such relationships in portfolio analysis, I turned to an expert from a consulting firm, ICF International, to review the DOE and NRC data and to characterize project relationships. A licensed Professional Engineer with 30 years of chemical industry experience, he examined each of the 171 pair-wise relationships among the 19 projects and classified it according to the taxonomy shown in Figure 4.2 (Lanza, 2012).

As shown in Table 4.3, the technological outcomes of 16 project pairs are likely to be correlated while Table 4.4 indicates that nine of the projects are likely to display a relationship in the marketplace. Four of the 16 pairs display an asymmetric technical relationship in which the

DOI: 10.1057/9781137542090.0008

TABLE 4.2 *Characterization of R&D projects*

| Projects | R&D project data | | | Potential project outcomes | | | |
Project number and name	Start year	End year	R&D cost (2003 $ million)	Date of deployment decision	Deployment cost (2003 $ million)	Probability of technical success	Probability of market acceptance
1 Ionic liquids	2004	2009	$3.02	2011	$4.98	0.20	0.55
2 Oxidation of alkanes	2004	2007	$6.29	2008	$8.37	0.06	0.26
3 Gas to ethylene	2004	2009	$1.79	2011	$14.42	0.07	0.18
4 Oxidation catalysts	2004	2010	$1.83	2013	$49.66	0.05	0.16
5 H$_2$O$_2$ production	2002	2008	$3.44	2009	$13.53	0.08	0.38
6 Catalytic hydrogenation	2003	2007	$2.95	2008	$0.70	0.13	0.27
8 Process intensification	2004	2008	$2.69	2009	$4.86	0.05	0.28
9 PTA purification	2002	2006	$3.20	2007	$0.38	0.28	0.20
10 Tackifier dispersants	2003	2006	$2.53	2007	$0.26	0.33	0.30
11 Efficient NGL removal	2003	2006	$3.28	2007	$25.04	0.18	0.17
12 Low VOC coatings	2004	2006	$4.78	2009	$0.65	0.18	0.52
13 Acetic acid	2003	2007	$2.84	2010	$9.36	0.20	0.17
14 Integrated distillation	2004	2008	$3.37	2009	$10.58	0.28	0.50
16 HC separation	2004	2006	$3.82	2007	$1.58	0.15	0.38
18 Chemical corrosion mgt	2003	2005	$1.80	2006	$4.60	0.50	0.67
19 Fluidized beds	2004	2007	$2.56	2009	$10.71	0.14	0.27
20 Production of paraxylene	2003	2006	$3.79	2019	$1.71	0.25	0.10
21 LPG recovery	2003	2003	$1.51	2004	$2.77	0.25	0.12
22 Rotary burner	2003	2005	$12.06	2006	$7.90	0.38	0.38

Projects #7, #15, and #17 were removed from the analysis, for reasons described in the text.

Source: Adapted from Linquiti (2012b).

Technical Relationships: "Each project pair was [characterized using a 7-point scale] with respect to its technological relationship during the R&D process, i.e., the extent, if any, to which the technological success of one project could inform the likelihood of technological success... of a second project. For example, if it is assumed that Project #1 is technologically successful, would that promote any meaningful inference concerning the likelihood of technological success of Project #2?

1. High Positive Correlation: If R&D Project #1 is technically successful, it is very likely that Project #2 will succeed (and vice versa).
2. Moderate Positive Correlation: If R&D Project #1 is technically successful, it is moderately likely that Project #2 will succeed (and vice versa).
3. Weak Positive Correlation: If R&D Project #1 is technically successful, it is somewhat likely that Project #2 will succeed (and vice versa).
4. No Correlation: The technical success of R&D Projects #1 and #2 are completely unrelated. Knowing the technical outcome of one project has no predictive power when considering the likelihood of technical success in the second project.
5. Weak Negative Correlation: If R&D Project #1 is technically successful, it is somewhat likely that Project #2 will fail (and vice versa).
6. Moderate Negative Correlation: If R&D Project #1 is technically successful, it is moderately likely certain that Project #2 will fail (and vice versa).
7. High Negative Correlation: If R&D Project #1 is technically successful, it is highly likely that Project #2 will fail (and vice versa)."

Commercial Relationships: "Each project pair was [characterized using a 7-point scale] with respect to its market relationship during deployment of the project. ...it was assumed that each project was technologically successful and was deployed in the marketplace, and the effect of such deployment on [the] other project was evaluated. For example, if ... Project #1 is successful in the marketplace, would that promote any meaningful inference concerning the likelihood of marketplace success of Project #2?

1. Strong Complements: The two technologies are highly related, and would very frequently be deployed in tandem at plants across industry.
2. Moderate Complements: The two technologies are related, and would often be deployed in tandem in at plants in many parts of industry.
3. Weak Complements: The two technologies are somewhat related, and would occasionally be deployed in tandem in plants in some areas of industry.
4. Strong Substitutes: The two technologies are apt to always be strong competitors to each other, and very rarely would be deployed in tandem at plants anywhere in industry.
5. Moderate Substitutes: The two technologies are apt to often be competitors to each other, and would likely not be deployed in tandem at most plants in the industry.
6. Weak Substitutes: The two technologies are apt to sometimes be competitors to each other, and would sometimes not be deployed in tandem at some plants in the industry.
7. Independent: ... technologies are unrelated, either in a technological sense or because they [apply to] *different sub-sectors* of the... industry, and the decision by a firm to deploy one of the technologies would have no bearing on the decision to deploy the second technology."

FIGURE 4.2   *Description of characterization of project relationships*

Source: Lanza (2012). Reproduced with permission.

DOI: 10.1057/9781137542090.0008

TABLE 4.3   *Technological relationships during R&D*

Pair		
Project A	Project B	Description of relationship*
1	8	Weak positive
2	3	Asymmetric—Weak positive/Moderate positive
2	4	Weak positive
2	5	Weak positive
2	6	Weak positive
3	4	Asymmetric—Moderate positive/Weak positive
3	5	Weak positive
3	6	Weak positive
4	5	Weak positive
4	6	Weak positive
5	6	High positive
5	14	Moderate positive
6	14	Moderate positive
11	20	Asymmetric—No correlation/Weak positive
11	21	Asymmetric—No correlation/Weak positive
20	21	Moderate positive

*When the relationship is asymmetric, the first characterization presented is the expected effect on Project B if Project A succeeds and the second characterization is the expected effect on Project A if Project B succeeds.

Source: Lanza (2012).

TABLE 4.4   *Commercial relationships during deployment*

Pair		
Project A	Project B	Description of relationship
2	4	Moderate substitutes
3	4	Strong substitutes
3	14	Strong complements
4	14	Strong complements
9	13	Weak complements
9	20	Strong complements
11	16	Strong substitutes
11	21	Weak complements
13	20	Weak complements

Source: Lanza (2012).

DOI: 10.1057/9781137542090.0008

implications of technical success in the first project for the success in the second are not the same as the converse situation. For example:

> Project 3 [is a] multi-step process [with] more than one principal technology. In order for Project 3 to be deemed technologically successful, all four of the principal technologies associated with the Project 3 multi-step process would need to be technologically successful. However, Project 4 [focuses on] only one principal technology, similar to one of the four principal technologies evaluated in Project 3. Therefore there is a potentially asymmetric technological relationship between Project 3 and Project 4.... Therefore the technological success of Project 3 would have a stronger influence on Project 4 than vice versa. (Lanza, 2012, p. 3)

In such cases, the asymmetric technical relationships in the pair were recorded.

4.2.2   Energy price data

Because the studied R&D projects focus on energy efficiency, energy prices are central to project valuation. The benefits of the studied R&D projects result from reductions in the use of one or more fuel types: petroleum distillate, natural gas, coal, and electricity. Some projects result in reduced use of a single fuel type; others create savings of multiple fuel types. While the NRC panel did consider environmental and national security benefits associated with reduced energy use, it did not monetize such benefits. I also value the R&D projects solely as a function of their energy savings. Three aspects of energy prices are relevant for this study—future prices, historical prices, and risk-neutral prices.

Future price projections

The NRC panel valued energy savings through the year 2030 based on a price forecast that was initially developed by the Energy Information Administration and then slightly adjusted by the panel. To maximize comparability with the NRC analysis (and having no reason to second-guess its price forecast), I use the same forecast of energy prices as did the panel. (As explained later, I do allow the price path to vary stochastically and adjust it for risk-neutrality.)

Table 4.5 shows the forecast price trajectories for the four fuel types. All prices are in 2003 dollars per million BTUs. In order to facilitate later analyzes, I also computed the year-over-year price changes, expressed as proportional returns, as well as the mean (μ) and standard deviation (σ) of those returns.

DOI: 10.1057/9781137542090.0008

TABLE 4.5 Future energy prices

Year	2003 $ per million BTUs				Year-over-year-change			
	Distillate	Natural gas	Coal	Electricity	Distillate	Natural gas	Coal	Electricity
2004	7.965	6.049	1.567	15.715				
2005	7.653	6.077	1.604	15.598	-0.0392	0.0046	0.0237	-0.0075
2006	7.153	5.429	1.594	15.055	-0.0654	-0.1066	-0.0060	-0.0348
2007	6.821	4.978	1.577	14.416	-0.0464	-0.0830	-0.0109	-0.0425
2008	6.878	4.610	1.578	14.123	0.0084	-0.0741	0.0006	-0.0203
2009	6.717	4.515	1.571	14.033	-0.0233	-0.0205	-0.0042	-0.0064
2010	6.783	4.373	1.565	13.842	0.0097	-0.0314	-0.0038	-0.0136
2011	7.134	4.372	1.564	13.699	0.0518	-0.0003	-0.0005	-0.0103
2012	7.160	4.444	1.555	13.848	0.0037	0.0164	-0.0060	0.0109
2013	7.188	4.541	1.547	14.089	0.0038	0.0220	-0.0048	0.0174
2014	7.176	4.701	1.548	14.438	-0.0016	0.0351	0.0004	0.0248
2015	7.194	4.818	1.550	14.622	0.0025	0.0249	0.0011	0.0128
2016	7.219	4.790	1.549	14.681	0.0035	-0.0057	-0.0001	0.0040
2017	7.277	4.823	1.557	14.817	0.0080	0.0069	0.0048	0.0093
2018	7.323	4.956	1.560	15.036	0.0063	0.0275	0.0018	0.0148
2019	7.379	5.115	1.561	15.313	0.0076	0.0321	0.0011	0.0184
2020	7.372	5.233	1.563	15.467	-0.0009	0.0231	0.0008	0.0101
2021	7.446	5.325	1.568	15.609	0.0100	0.0176	0.0036	0.0092
2022	7.515	5.356	1.571	15.623	0.0092	0.0058	0.0018	0.0009
2023	7.495	5.350	1.604	15.607	-0.0025	-0.0011	0.0208	-0.0010
2024	7.563	5.408	1.605	15.663	0.0091	0.0108	0.0010	0.0036
2025	7.733	5.472	1.602	15.752	0.0224	0.0119	-0.0017	0.0057
2026	7.834	5.537	1.591	15.814	0.0131	0.0118	-0.0070	0.0040
2027	7.936	5.602	1.580	15.877	0.0131	0.0118	-0.0070	0.0040
2028	8.040	5.668	1.569	15.940	0.0131	0.0118	-0.0070	0.0040
2029	8.145	5.735	1.558	16.004	0.0131	0.0118	-0.0070	0.0040
2030	8.251	5.803	1.547	16.067	0.0131	0.0118	-0.0070	0.0040
					$\mu = 0.0016$	$\mu = -0.0010$	$\mu = -0.0004$	$\mu = -0.0010$
					$\sigma = 0.0229$	$\sigma = 0.0355$	$\sigma = 0.0078$	$\sigma = 0.0154$

Source: NRC data (Fri, November 2009).

DOI: 10.1057/9781137542090.0008

Historical price analysis

The NRC panel did not evaluate historical energy prices, but they are characterized here for three reasons. First, one of the ways in which the 19 R&D projects may exhibit interdependence in their commercial performance is through variation in energy prices. If, for example, natural gas prices rise, but coal prices fall, those technologies that reduce gas use will become more valuable while, at the same time, those that improve the efficiency of coal use will lose value. Characterizing the co-variations of energy prices is thus essential to the analysis; I chose to examine historical patterns in order to characterize these relationships. Second, simulation of an uncertain, or stochastic, set of future prices requires a measure of year over year price variability. I make the assumption that the standard deviation of historical energy price changes is an appropriate proxy for this variability and examine the historical price data to obtain it. Finally, to develop the risk-neutral price path described below, I need to relate changes in energy prices to the returns on a broad market basket of risky financial investments; doing so is a backward-looking, historical, exercise.

Historical prices for the four relevant fuel types were taken from multiple reports issued by the EIA. Data were available for an 11-year period and are displayed in Table 4.6. In addition to nominal prices per million BTUs, the year-over-year changes are calculated and the mean and standard deviation computed. The correlation coefficients among the annual returns are presented in Table 4.7.

Risk-neutral pricing

As explained previously, two methods are available for valuing real options: the no-arbitrage, or replicating portfolio, approach and the risk-neutral approach. Because R&D projects are typically not traded in the marketplace (making it impossible to form a replicating portfolio), this analysis employs the risk-neutral approach. Doing so requires that we "risk-adjust" either the probabilities or the prices used in the analysis (Triantis, 2003). I take the latter approach in this analysis.

I begin with the premise that the energy price forecasts in Table 4.5 contain an embedded risk premium. My objective is thus to compute the premium and remove it from the price forecasts. At first glance, it may not appear that the EIA/NRC price forecast includes a risk premium; they are, after all, the best judgments of the energy market experts at EIA about likely future prices of energy. As these analysts established the

TABLE 4.6 *Historical energy prices*

	Nominal $ per million BTUs				Year-over-year change			
Year	Distillate	Natural gas	Coal	Electricity	Distillate	Natural gas	Coal	Electricity
1992	4.87	3.07	1.47	14.87				
1993	4.78	2.79	1.45	14.53	-0.018	-0.091	-0.014	-0.023
1994	4.59	2.56	1.45	14.75	-0.040	-0.082	0.000	0.015
1995	4.61	2.28	1.48	14.54	0.004	-0.109	0.021	-0.014
1996	5.50	2.96	1.46	13.54	0.193	0.298	-0.014	-0.069
1997	5.10	2.96	1.44	13.56	-0.073	0.000	-0.014	0.001
1998	4.02	2.66	1.45	13.09	-0.212	-0.101	0.007	-0.035
1999	4.65	2.79	1.43	13.09	0.157	0.049	-0.014	0.000
2000	7.21	4.31	1.41	13.50	0.551	0.545	-0.014	0.031
2001	6.55	4.87	1.46	14.10	-0.092	0.130	0.035	0.044
2002	6.21	3.75	1.52	14.74	-0.052	-0.230	0.041	0.045
					$\mu = 0.042$	$\mu = 0.041$	$\mu = 0.004$	$\mu = 0.000$
					$\sigma = 0.214$	$\sigma = 0.231$	$\sigma = 0.022$	$\sigma = 0.036$

Sources: EIA (January 1994, January 1995, January 1996, December 1996, December 1997, December 1998, December 1999, December 2000, December 2001, January 2003, January 2004).

TABLE 4.7   *Correlations among historical energy price changes*

	Distillate	Natural gas	Coal	Electricity
Distillate	1.0000			
Natural Gas	0.8439	1.0000		
Coal	−0.4669	−0.4517	1.0000	
Electricity	0.0665	−0.0277	0.5207	1.0000

Source: Linquiti (2012b).

yearly price estimate for each fuel type, however, they had no choice but to take into account myriad risks related to the future state of the global macroeconomy, the evolving political economy of energy pricing, and the intrinsic drivers of energy supply and demand. These are referred to as systematic risks that investors are unable to diversify away by, for example, increasing the number of their holdings (Brealey & Myers, 2003). These price forecasts thus contain a risk premium that must be removed before the price series can be considered risk-neutral.

To do so, I apply the Capital Asset Pricing Model (CAPM). While the CAPM is recognized to have flaws, I chose to use the CAPM because it provides "a starting point for thinking about the risk-return relationship" (Bodie, Kane, & Marcus, 2009, p. 299). In the CAPM framework, the expected risk premium (ERP$_s$) on stock S is equal to the beta of the stock times the expected risk premium observed for the market as whole, or

$$ERP_s = R_s - R_f = \beta_s \cdot (R_m - R_f) \tag{4}$$

where R_s, R_f, and R_m are the percentage return on, respectively, the stock, a risk-free bond, and the broad market, and β_s represents the relationship between the returns on stock, S, and the market, M (Brealey & Myers, 2003, p. 195).

Beta is computed by regressing the percentage returns of a stock on the percentage returns of the market. Betas greater than one imply that the stock's movements are larger than those of the market while a beta of zero implies that the movement of the stock return and the market return are unrelated. Betas of less than zero imply that the stock's returns tend to move opposite those of the market. In this case, the energy fuel type is analogous to the stock, and I treat the year-over-year price changes as the return. Table 4.8 shows the derivation of beta for each of the four fuel types. As a proxy for the equity market as a whole, I used the S&P500 index and computed both its average level over each year and its annual return on a year-over-year basis. The estimated betas for

DOI: 10.1057/9781137542090.0008

TABLE 4.8   *Derivation of beta by fuel type*

Fuel type	Intercept			Beta			
	Coefficient	Std error	p-value	Coefficient	Std error	p-value	R^2
Distillate fuel	0.0304	0.0853	0.7306	0.1108	0.4518	0.8125	0.0075
Natural gas	0.0223	0.0916	0.8141	0.1790	0.4856	0.7220	0.0167
Coal	0.0134	0.0060	0.0553	−0.0949	0.0316	0.0170	0.5301
Electricity	0.0166	0.0097	0.1251	−0.1633	0.0513	0.0130	0.5586

Source: Linquiti (2012b).

coal and electricity are statistically significant at the 0.02 level. Given the standard errors associated with the estimated betas for distillate fuel and natural gas, however, we cannot statistically distinguish these two betas from zero. Nonetheless, in the absence of a better estimate of beta, I rely on the reported beta for subsequent analysis.

A 2003 review of the historical record estimated a forward-looking equity risk premium of 6 percent relative to long-term government bonds (Ibbotson & Chen, 2003). I assume a risk-free rate of 3 percent (U.S. Office of Management and Budget, 2003).

With a beta for each fuel type, and a market risk premium of 6 percent, I estimate the risk premium for each fuel source. This premium is then subtracted from the arithmetic average of the forecast fuel price changes between 2004 and 2030 (shown in Table 4.5) to compute the risk-neutral growth rate of energy prices. Volatility is assumed to be the same in both the risk-neutral world and the risk-laden world, and thus does not require adjustment (Hull, 2009). When simulating future energy prices, the four price paths (one for each fuel type) are correlated using the historical values from Table 4.7. Table 4.9 provides the risk-neutral price path (characterized by the mean return and standard deviation) for each fuel type.

4.3   Methodology

This section summarizes how the data described in the prior section have been analyzed. I first describe how NRC's analysis is replicated and extended. I then turn to a portfolio characterization that frames each project as a real option and explicitly analyzes relationships among them. I next explain a series of sensitivity analyses and then present my approach to measuring portfolio risk.

DOI: 10.1057/9781137542090.0008

TABLE 4.9   *Estimation of risk-neutral price path*

Fuel type	Risk-premia			Price paths		
	Beta	Market risk premium	Fuel type risk premium	Forecast price growth	Risk-neutral growth rate (μ)	Std dev (σ) of historical returns
Distillate fuel	0.1108	0.0600	0.0066	0.0016	−0.0050	0.2136
Natural gas	0.1790	0.0600	0.0107	−0.0010	−0.0117	0.2306
Coal	−0.0949	0.0600	−0.0057	−0.0004	0.0052	0.0217
Electricity	−0.1633	0.0600	−0.0098	0.0010	0.0108	0.0364

Source: Linquiti (2012b).

4.3.1   Replicate and extend NRC methodology

As shown in Figure 4.1, the NRC panel provided only a graphic depiction of the distribution of potential portfolio results and did not provide either a qualitative or a quantitative description of that distribution. Accordingly, in order to set a comparative benchmark for the results obtained by my additional analysis, I used a Monte Carlo model to replicate the NRC analysis while simultaneously computing the statistics necessary to describe the resulting probability distribution of portfolio values. To preserve consistency with the portfolio analysis I present later, I omitted the same three projects that had been dropped from the data set for the reasons outlined above and analyze only the remaining 19 projects. Like the NRC panel, I examined only the benefits of each project and did not subtract its R&D cost. In addition to computing the expected benefit of each project, I computed summary statistics for the probability distribution of portfolio values.

4.3.2   Execute real option analysis with related projects

This analysis uses a second Monte Carlo model both to compute option values and to simulate portfolio performance. Each of the 250,000 iterations of the model represents a complete simulation of the entire portfolio's performance over the full time horizon.

Simulation of energy prices

The prices of the four energy fuel types from 2003 to 2030 are simulated once per iteration. Doing so ensures that all projects are valued using the same energy prices. Energy prices are assumed to follow a risk-neutral

DOI: 10.1057/9781137542090.0008

random walk, based on geometric Brownian motion, with a mean and standard deviation equal to the values shown in Table 4.9. Correlations among the price path taken by each fuel type are simulated using the values in Table 4.7. Fluctuations in energy prices have the practical effect of introducing a relationship among the value of the R&D projects.

Simulation of project-level results

Once the energy prices have been simulated, the model moves to a project-by-project analysis. The technical success or failure of each project is simulated first, based on the probabilities shown in Table 4.2. For the 16 project pairs deemed to be technically correlated, the correlations displayed in Table 4.3 are applied, with weak, moderate, and high correlations assumed to correspond to values of 0.25, 0.50, and 0.75, respectively. For the four projects understood to have asymmetric relationships, the correlation coefficient was set at the mid-point of the two relevant correlations. For example, the correlation between the success of Projects 2 and 3, deemed to be weak positive in one direction, and moderate positive in the other, was set at 0.38, or halfway between 0.25 (weak) and 0.50 (moderate).

If a project is simulated as successful, its potential energy savings are monetized on the basis of the NRC panel's estimate of physical fuel use savings, and the set of energy price paths simulated for the current model iteration. Next, the model moves on to evaluate potential market relationships in the nine project pairs shown in Table 4.4 to have either a complementary or substitute relationship. If either or both of the two projects have encountered technical failure, no further analysis of project relationships is done.

If, however, both projects are simulated as technically successful, then the model adjusts the expected commercial value of each. While the adjustment is arbitrary, the set of adjustments is internally consistent and meant to be a proxy for types of effects that might be seen during deployment in the marketplace. If the two projects are complements, then the value of the energy cost savings for both projects is increased by 5 percent, 10 percent, or 15 percent, depending on whether they are, respectively, weak, moderate, or strong complements. The effect of the adjustment is to increase the likelihood of both technologies being deployed in the market. If the two projects are deemed to be substitutes, the model determines which of the two has the higher cost savings (in absolute terms). The model then simulates the strengthening of that technology's market position by increasing the value of its energy savings by 5 percent, 10 percent, or 15 percent depending whether it is, respectively, a weak, moderate, or strong substitute. Conversely, the value of the energy savings from the second, intrinsically less valuable, project is decremented by 5 percent, 10 percent, or 15 percent; because they are substitutes in the market, the success of the first technology reduces the value of the second. This approach is clearly a simplification of a complex reality, but has been included so that we can assess the implications of incorporating or ignoring potential project relationships during portfolio valuation.

Computation of option values by project

I treat each R&D project as a European call option and value it as of 2002. Even though some projects start in 2003 or 2004, I make the assumption that the portfolio is being formed in 2002 and that all the options therefore are being valued in 2002. To compute the option value, I start with the deployment decision date shown in Table 4.2 and treat it as the expiration date of the option. The present value of all future energy cost savings are computed, using the risk-neutral price paths for the relevant fuel types, and discounted at the risk-free rate of 3 percent back to the year of option expiration.

The model compares the value of the future savings to the exercise price (i.e., the technology deployment cost as shown in Table 4.2) to determine whether the option is exercised (i.e., the new more energy-efficient technology is deployed). The value of the option is then computed as the maximum of zero (i.e., the case where the deployment costs exceed the value of the energy savings) and the energy cost savings less the deployment cost. This value is discounted back to 2002, using the risk-free rate, to characterize the value of the project as of the date of portfolio formation. This comparison of energy cost savings to deployment cost is the only driver of whether the results of a technically successful R&D project are deployed in the marketplace. As such, my approach does not use the probabilities of market acceptance of each technology that were estimated by the NRC panel.

Simulation of portfolio characteristics

After the 19 R&D projects have been valued as real options, the model computes the portfolio value by summing all project values. The model then moves to the next iteration, first simulating another set of risk-

DOI: 10.1057/9781137542090.0008

neutral energy price paths, then executing the project-by-project option analysis, and finally computing the portfolio value. After 250,000 iterations have been completed, the model computes summary statistics that can be used to analyze the return and the risk of the portfolio. An option value for each project is also computed, based on the average option value across all iterations.

4.3.3 Sensitivity analyses

In order to better understand the implications of a failure to incorporate relationships among projects when conducting portfolio analyses, two types of sensitivity analyses are conducted. First, the simulation described above is repeated without including any project relationships. To that end, the correlations between the energy price paths, between the 16 project pairs deemed to technically related, and between the 9 project pairs likely to be related in the marketplace are all set to zero. The results thus reflect only the independent operation of the relevant variables.

The second sensitivity analysis probes the impact of different types of correlations among projects, from strongly positive to negative. In these cases, the project relationships ignore the expert-estimation of technical and market relationships, and instead rely on assumed correlations among the probabilities of technical success that vary from 1.0 to 0.0, in increments of 0.25. In addition, in one scenario, projects are ranked on the basis of expected payoff and then paired (i.e., the first and second project are paired, the third and fourth project are paired, and so on). Each project pair is then assigned a correlation coefficient of −1.0.

In this second set of sensitivity analyses, the correlations among energy prices *are* simulated. My rationale for doing so is that such correlations originate in the marketplace and are not subject to control by the R&D policymaker, and would persist irrespective of the set of R&D projects placed in the portfolio by that policymaker.

4.3.4 Measuring portfolio risk

I, and others, have argued that government managers making R&D investment decisions should always be concerned with both return *and* risk (Ruegg & Feller, 2003; Linquiti, 2012a; Tassey, 2003). Doing so is common practice in the private-sector where "casual observation and formal research both suggest that investment risk is as important to investors as expected return" (Bodie, Kane, & Marcus, 2009, p. 113).

DOI: 10.1057/9781137542090.0008

In other words, investors care not only about an investment's likely return, but also the risk of variations in the return. In statistical terms, the argument here is that the mean of the probability distribution of likely portfolio returns provides an incomplete picture of the investment and must be supplemented with measure(s) of the dispersion in the distribution.

One measure of risk is the standard deviation of potential outcomes (Bodie, Kane, & Marcus, 2009). Because dissimilar distributions can have the same standard deviation, however, the standard deviation only provides an unambiguous characterization of risk when the investments under consideration have the same underlying distribution (e.g., the normal distribution). As explained in Section 3.2, however, the payoffs to real options are not symmetrically distributed, as the optionality truncates downside risk. When options are combined into portfolios, it is even harder to predict the nature of the distribution of possible outcomes. Hence, while the standard deviation of portfolio returns can inform considerations of risk, it will not tell the whole story. Instead, Tassey argues that "risk is the probability of an outcome below a minimum acceptable level" (Tassey, 1997, p. 83).

For purposes of this analysis, I operationalize Tassey's advice by computing three different thresholds for portfolio returns and assess risk based on the probability that each threshold will not be achieved. The first threshold is set at $67.54 million, which is the total investment in the 19 R&D projects. The risk, in this case, is that the aggregate payoff from the 19 projects will not even exceed the funds expended to conduct them. The other two thresholds are based on benchmarks suggested by Tassey. He reports that corporate R&D managers typically expect an annual return of 20–25 percent and also argues that government research projects ought to aim for an annual return of 50 percent (Tassey, 2003, pp. 35–36). To construct the second and third threshold, I thus used expected returns of 25 and 50 percent, adjusted to factor out inflation, which averaged 5 percent per year between 1972 and 2003 (U.S. Office of Management and Budget, 2011, B-64). This computation yielded a second threshold of $241.40 million which would be the return on the 19 R&D projects if the corporate R&D hurdle rate were attained and a third threshold at $543.14 million which would be the return if Tassey's suggested nominal 50 percent threshold were achieved.

DOI: 10.1057/9781137542090.0008

4.4   Results

Results are presented in three parts below. The first section focuses on project valuation while the second addresses portfolio valuation and presents the results of applying a real option approach and incorporating interdependencies among projects in the portfolio. The final section summarizes two sensitivity analyses.

4.4.1   Project valuation

Table 4.10 presents two estimates of the value of each project, net of R&D costs. The first estimate uses NRC's DT/DCF approach and a 3 percent discount rate while the second uses a real options approach. Net project values range from a loss of $3.54 million to a gain of $279.96 million, depending on the project and the valuation method.

In every case, the real option methodology yields a higher valuation than the DT/DCF approach. In the aggregate, real option values are collectively $835 million higher than DT/DCF values. The key driver of

TABLE 4.10   *Project valuation net of R&D cost (2003 $million)*

	Valuation method	
Project name	DT/DCF @3%	Real option
Ionic liquids	16.91	49.23
Oxidation of alkanes	18.12	51.13
Gas to ethylene	10.76	83.66
Oxidation catalysts	17.44	165.64
H_2O_2 production	43.77	130.90
Catalytic hydrogenation	4.28	16.35
Process intensification	1.63	14.04
PTA purification	0.41	12.16
Tackifier dispersants	−0.43	2.26
Efficient NGL removal	32.08	169.16
Low VOC coatings	−3.09	−2.87
Acetic acid	21.79	108.76
Integrated distillation	217.38	279.96
HC separation	13.59	19.42
Chemical corrosion mgt	54.52	89.63
Fluidized beds	13.56	47.99
Production of paraxylene	−3.54	0.79
LPG recovery	2.51	22.05
Rotary burner	41.89	77.95
Total value	503.61	1,338.21
Difference	834.60	

DOI: 10.1057/9781137542090.0008

the differences is that the real option method simultaneously captures the virtually unlimited upside of each project (i.e., the value of energy efficiency under conditions of extremely high energy prices) while also bounding the downside risk (i.e., the maximum loss can never exceed amount of money spent on the R&D project).

Another interesting result is that the DT/DCF and real options methods would lead a policymaker to prioritize R&D projects in different ways. Figure 4.3 shows the differences in the rankings that result from the two valuation methods. Each mark on the graph represents one project; if the mark does not fall on the line, then the project does not have the same rank under the two methods—the case with 14 of the 19 projects. For example, the second most valuable project under the DT/ DCF method drops to sixth place under the real options approach while the 12th ranked project climbs to seventh place.

The Tackifier Dispersants project and the Paraxylene project both demonstrate another interesting consequence of the choice of valuation method. Both projects display a negative value under the DT/DCF method, suggesting that all else equal, they are not a good use of public funds, but when seen as real options, the projects show a positive value,

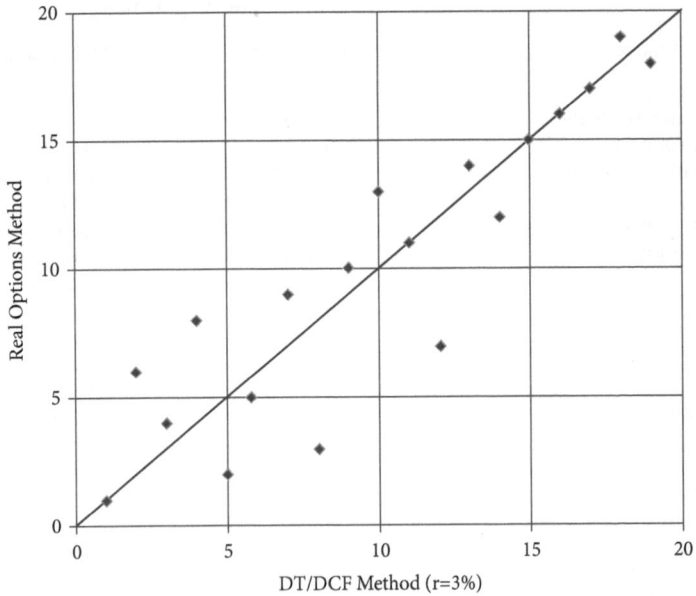

FIGURE 4.3 *Comparison of project rankings based on valuation method*
Source: Adapted from Linquiti (2012b)

DOI: 10.1057/9781137542090.0008

albeit a small one, indicating the opposite conclusion (i.e., that policy-makers should view these two projects as socially beneficial).

4.4.2 Portfolio valuation

Shifting our focus from individual R&D projects to the portfolio level further underscores the impact of applying a real option valuation instead of the more traditional DT/DCF method. To re-cap, the primary differences in my analysis of the two methods are that, first, with the real option method, I allow energy savings (i.e., project benefits) to vary stochastically and simulate the deployment decision as a real option choice based on a comparison of the value of energy cost savings and technology deployment costs. Second, unlike the approach taken by NRC, I simulate the portfolio with interdependencies—technical and commercial—among several of the R&D projects. Table 4.11 statistically summarizes the distribution of portfolio values (without subtracting R&D costs) generated by the two methods, while Figure 4.4 presents the two distributions in graphic form.

TABLE 4.11 *Simulation of portfolio value: decision tree/discounted cash flow method versus real option method (2003$ million)*

250,000 Iterations

Statistic	DT/DCF method ($r = 3\%$) and uncorrelated projects	Real option method with expert-estimation of project correlations
Mean	$570.9	$1,405.8
Standard deviation	$750.7	$1,980.8
Coefficient of variability	1.31	1.41
0%	$0.0	$0.0
10%	$0.0	$154.1
20%	$0.0	$261.5
30%	$39.5	$389.2
40%	$166.9	$535.6
50%	$188.4	$732.5
60%	$373.6	$1,019.0
70%	$540.5	$1,449.0
80%	$1,336.7	$2,116.4
90%	$1,743.6	$3,433.7
100%	$6,996.3	$132,262.1
Prob(Portfolio value < $67.54m)	31.6%	6.0%
Prob(Portfolio value < $241.40m)	53.7%	18.2%
Prob(Portfolio value < $543.14m)	70.3%	40.4%

Difference of means (unequal variance): $t = -200.0$, $p = 0.0000$
Difference of standard deviations (robust): Levine's $F_{(1, 499998)} = 36794.0$, $p = 0.0000$

DOI: 10.1057/9781137542090.0008

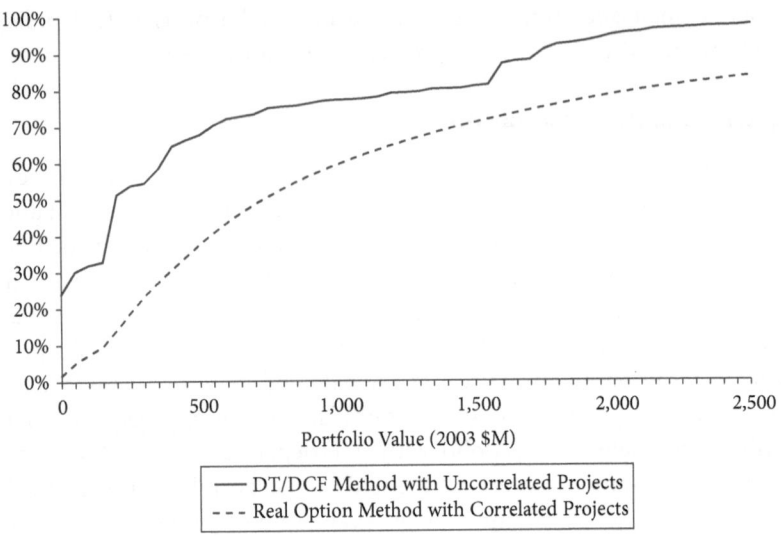

FIGURE 4.4   *Portfolio value: decision tree/discounted cash flow method versus real option method*

Both the mean value and the standard deviation are significantly higher under the real option method and statistical testing suggests that there is virtually no chance that these differences result from random chance. More specifically, the mean value of the portfolio is estimated at $571 million with the DT/DCF method, while the real option method suggests a value of just over $1.4 billion.[6] Similarly, the standard deviation increases from $751 million to $1.98 billion, although because of the difference in the mean values, the standard deviations are not directly comparable. Instead, the coefficient of variability (i.e., a unitless measure in which the standard deviation divided by the mean) is more informative, increasing from 1.31 to 1.41.

Whether the portfolio is seen as more risky with one method or the other is a matter of interpretation. If one subscribes to the traditional view that the standard deviation measures risk, then the real option approach suggests a higher risk owing to the increase in the coefficient of variability relative to the DT/DCF method. If, however, one subscribes to the view (suggested by Tassey) that risk is the likelihood of failing to achieve a benchmark level of return, then the DT/DCF method indicates a much higher risk. For example, the DT/DCF method points toward a

DOI: 10.1057/9781137542090.0008

31.6 percent chance of failing to create benefits sufficiently large to pay back the initial investment in the R&D while the real options method suggests that there is only a 6.0 percent chance of such an outcome. Similarly, if the policymaker aims to achieve the nominal 50 percent return proposed by Tassey, then there is approximately a 70 percent chance of failure under the DT/DCF method, but just a 40 percent chance of failure under the real options method.

Figure 4.4 plots cumulative density plots for both distributions of portfolio values. For any potential portfolio value, there is a higher likelihood that it will not be achieved when using the DT/DCF method rather than the real option method. In other words, the curve for the latter distribution is always below the curve for the DT/DCF approach. Indeed, the difference at the vertical intercept is stark: the DT/DCF method suggests about a 24 percent chance of receiving no benefit at all from the R&D investment while the real option method suggests that this risk is closer to 2 percent.

4.4.3   Sensitivity analyses

The first sensitivity analysis relaxed the assumption that the technical and commercial performance of the projects in the portfolio would be interdependent, while still applying the real option methods. To do so, I assumed that none of the projects had a technical or commercial relationship with any other project, and that the energy price paths for the four fuels were uncorrelated. Results are presented in Table 4.12 and Figure 4.5.

When it comes to the effect of incorporating project relationships in the portfolio valuation process, we first observe that the portfolio value goes up somewhat (from $1.367 billion to $1.406 billion). Simulation of a commercial relationship between two projects is accomplished by raising the value of both projects if they are complements or, if they are substitutes, by raising the value of the project with the higher energy cost savings and reducing (by the same proportion) the value of the other project. In the aggregate, such adjustments are positive and raise the portfolio value compared to the case in which project relationships are not simulated.

The incorporation of project relationships also increases the apparent risk of the portfolio, irrespective of how one defines risk. A focus on the coefficient of variability shows an increase in risk (from 1.19 to 1.41). So

TABLE 4.12   *Sensitivity analysis: real option valuation with and without project correlations (2003$ million)*

250,000 Iterations

Statistic	Real option method with expert-estimation of project correlations	Real option method with assumption of no correlations among projects
Mean	$1,405.8	$1,367.4
Standard deviation	$1,980.8	$1,629.7
Coefficient of variability	1.41	1.19
0%	$0.0	$0.0
10%	$154.1	$176.7
20%	$261.5	$300.1
30%	$389.2	$446.5
40%	$535.6	$612.4
50%	$732.5	$834.9
60%	$1,019.0	$1,142.6
70%	$1,449.0	$1,543.2
80%	$2,116.4	$2,115.4
90%	$3,433.7	$3,209.3
100%	$132,262.1	$66,945.0
Prob(Portfolio value < $67.54m)	6.0%	5.0%
Prob(Portfolio value < $241.40m)	18.2%	15.4%
Prob(Portfolio value < $543.14m)	40.4%	36.1%

Difference of means (unequal variance): $t = 7.4782$, $p = 0.0000$
Difference of standard deviations (robust): Levine's $F(1, 499998) = 1297.1$, $p = 0.0000$

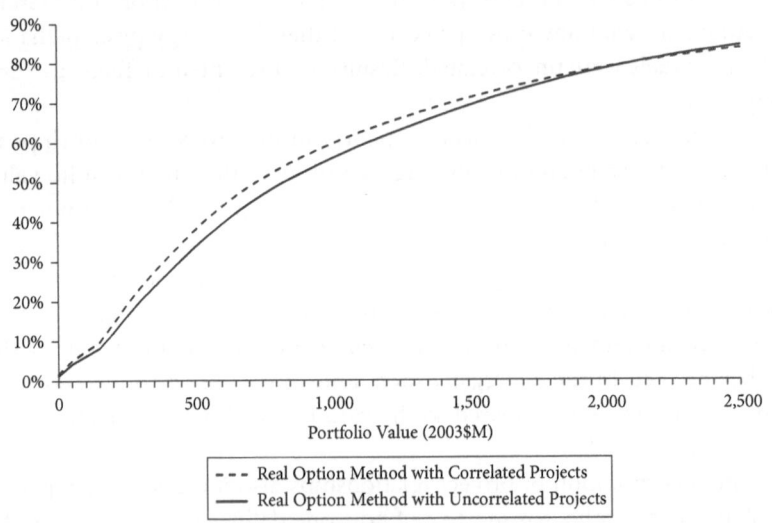

FIGURE 4.5   *Portfolio value: real option method with and without project correlations*

DOI: 10.1057/9781137542090.0008

too does a focus on the probability of failing to achieve a hurdle rate (e.g., for the breakeven hurdle, a 6.0 vs 5.0 percent chance of failure, and for the highest hurdle, 40.4 vs 36.1 percent). This result is attributable to the fact that on balance, the set of simulated relationships among projects is positive. All of the relationships among the 16 technically related project pairs exhibit a positive correlation. In addition, only three of the nine projects expected to be related in marketplace deployment are substitutes; the rest are complements, meaning that the value of commercial success of both is increased relative to a case where they are assumed to be unrelated.

In short, when project relationships are ignored (as they are in this sensitivity case), the analyst is implicitly assuming that the performance of all projects will be uncorrelated (i.e., that all the inter-project correlation coefficients are 0.00). If in reality, however, the projects are positively correlated in some way (as they are in the primary analysis here), then the failure to incorporate these relationships in the portfolio valuation will lead the analyst to underestimate the degree of risk associated with the R&D investment.[7]

The second sensitivity analysis examined the effect of alternative correlation structures on portfolio valuation. For this sensitivity, the expert-estimated project relationships were discarded. Instead, I assumed that each project was positively correlated with all other projects in the portfolio. I created five sensitivity scenarios by varying the positive correlation from 1.00 to 0.00 in increments of 0.25. Because it is mathematically impossible for a large group of projects to all be negatively correlated with one another, I created a sixth sensitivity scenario by forming nine pairs of projects with the two projects in each pair being of roughly similar expected value, and then set the correlation for each pair to –1.00.

I also assume, for all six scenarios, that the price paths for the four types of energy were correlated in the same way as in my primary analysis. My thinking here was that while policymakers can affect within-portfolio project correlations when they select specific R&D projects for the portfolio, they have no control over the energy prices that create interdependencies in the value of R&D outcomes irrespective of the policymaker's preferences. Results are presented in Table 4.13 and Figure 4.6.

As suggested by finance theory, the mean value of the portfolio is unaffected by the correlations among the projects in it. In all cases, the mean is about $1.377 billion. A standard F-test of multiple means, with

TABLE 4.13 *Sensitivity analysis: effect of assumed project correlation on real option valuation (2003$ million)*

250,000 Iterations

Statistic	Correlation of all projects = 1.0	Correlation of all projects = 0.75	Correlation of all projects = 0.50	Correlation of all projects = 0.25	Correlation of all projects = 0.00	Correlation of −1.00 for selected project pairs
Mean	$1,373.5	$1,383.7	$1,376.0	$1,379.3	$1,372.7	$1,375.1
Standard deviation	$3,481.8	$3,094.0	$2,592.6	$2,212.0	$1,750.6	$1,703.1
Coefficient of variability	2.54	2.24	1.88	1.60	1.28	1.24
0%	$0.0	$0.0	$0.0	$0.0	$0.0	$0.0
10%	$0.0	$0.0	$0.0	$0.0	$171.6	$199.4
20%	$0.0	$0.0	$0.0	$127.9	$288.5	$323.0
30%	$0.0	$0.0	$87.2	$231.7	$425.7	$458.3
40%	$0.0	$40.7	$212.0	$382.4	$583.2	$610.5
50%	$111.6	$201.8	$382.7	$593.9	$798.8	$823.2
60%	$204.3	$413.9	$681.7	$919.6	$1,107.5	$1,124.8
70%	$457.1	$890.6	$1,209.4	$1,422.5	$1,514.5	$1,522.3
80%	$1,516.8	$1,914.8	$2,115.8	$2,172.1	$2,097.7	$2,093.1
90%	$4,499.6	$4,286.1	$3,960.8	$3,624.0	$3,238.5	$3,195.9
100%	$159,772.6	$143,509.1	$105,399.9	$99,270.3	$76,326.0	$112,778.1
Prob(Portfolio value < $67.54m)	50.0%	40.9%	29.2%	17.2%	5.1%	3.6%
Prob(Portfolio value < $241.40m)	63.2%	53.3%	42.6%	30.8%	16.1%	13.5%
Prob(Portfolio value < $543.14m)	71.5%	63.5%	56.0%	47.9%	37.7%	36.0%

Difference of means test: $F_{(5,1499994)} = 0.66$, $p = .6545$
Difference of standard deviations test (robust): Levine's $F_{(5,1499994)} = 6913.2$, $p = .0000$

DOI: 10.1057/9781137542090.0008

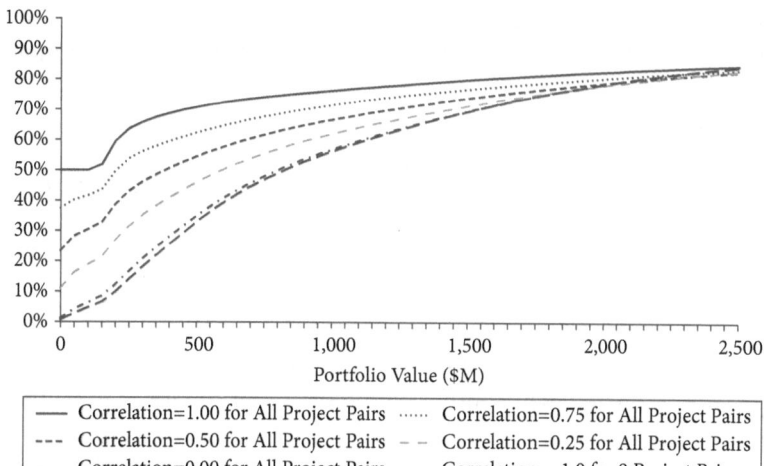

FIGURE 4.6   *Portfolio value: effect of project interdependencies*

$p = 0.6545$, gives us no reason to suspect that the slight differences among scenarios are statistically significant.

The standard deviations, however, vary widely with the highest being roughly twice the smallest. Levine's robust F-test for multiple standard deviations, with a $p = 0.0000$, suggests that it is statistically extremely unlikely that these variations in standard deviation are the result of random chance. As the correlations move in the negative direction, the standard deviations decrease. This general result is expected, based on standard finance theory, but its magnitude for a specific portfolio can only be ascertained through simulation.

The payoff to an option (real or financial) is not symmetric, being unbounded on the upside, but truncated at the option purchase price on the downside. This asymmetry has a significant limiting influence on the risk reduction, or hedging effect, that can be created by including negatively correlated projects. The two final scenarios—one with no project correlations at all and another with strong negative correlations among nine pairs of similarly sized projects—show only a relatively modest difference in standard deviation of $1.75 billion versus $1.70 billion, despite profoundly different correlation structures.

Shifting our focus from the standard deviation as a measure of risk to Tassey's framework in which risk is the likelihood of failing to achieve a target return, we see a similar pattern. As the projects in the portfolio

DOI: 10.1057/9781137542090.0008

become progressively less positively correlated, the risk of failure drops appreciably. Again, however, the risk reduction that occurs when moving from uncorrelated projects to a set of strong negative correlations among nine project pairs is relatively small, varying only by 1.5–2.6 percentage points.

In addition, with the 19 studied R&D projects focused on efficiencies for only four fuel types, it is inevitable that potential fluctuations in energy prices create a relationship among the potential commercial value of the R&D projects. The impact of energy price correlations can be isolated by comparing two of the scenarios presented above. The rightmost column of Table 4.12 provides portfolio statistics where no correlations of any type are simulated, while fifth scenario of the sensitivity analysis shown in Table 4.13 is one in which only energy price correlations are simulated. "Deep correlations," caused by an independent third variable, thus are the only driver of differences in these two cases.

The portfolio value is about the same (i.e., $1.37 billion), but the risk increases, albeit by a modest amount, when energy price correlations are introduced. The coefficient of variability rises from 1.19 to 1.28, the risk of failing to achieve a nominal 25 percent return (i.e., a portfolio value of $241.10 million) rises from 15.4 percent to 16.1 percent, and the risk of failing to achieve a nominal 50 percent return (i.e., a portfolio value of $543.14 million) rises from 36.1 percent to 37.7 percent. In short, while the more visible technical and commercial relationships among R&D projects affect portfolio risk, these "deep correlations," which may be harder to detect, can also be important.

Notes

1   "Prospective" simply means a valuation done prior to the start of the R&D project.
2   Some of the material in this chapter is drawn from my dissertation (Linquiti, 2012b).
3   See Section 2.6 for a more detailed discussion of this issue.
4   This is likely a strong assumption on NRC's part. Delays in the development of one technology may enhance the prospects of success for competing technologies that are developed in the interim.
5   It is not the estimated negative savings per se that makes these two projects impossible to analyze. Indeed, option analysis is well suited to such cases because it can capture the possibility of a positive payoff to the option, even

when the expected value of the payoff is negative (i.e., the option is currently out of the money). NRC estimated the cost of technology deployment (i.e., the option exercise price) as equal to three years' worth of projected energy savings. Given negative savings, the exercise price for these two projects would be negative—an empirical impossibility.

6 NRC reported a value of $534 million for the portfolio when simulated in a Monte Carlo model. Two factors contribute to the difference from the $571 million reported above. As noted previously, I dropped three projects from the analysis, two with projected net costs (rather than benefits) and one because of inconsistencies in its reported timing. Deletion of the two projects with costs, and the fact that NRC executed 1,000 iterations while I simulated 250,000, account for most of this difference.

7 The converse is also true. If the projects are actually negatively correlated, but the analyst ignores correlations, then portfolio risk would be overestimated.

DOI: 10.1057/9781137542090.0008

5

Recommendations for Public Sector R&D Managers

Abstract: *Chapter 5 provides a series of recommendations for public sector R&D managers based both on theoretical analysis and on a practical case study of a portfolio of "real world" R&D projects. In brief, discounted cash flow techniques should be abandoned in favor of real option methods. In addition, individual R&D projects should never be valued in isolation, but only in concert with the other R&D projects comprising a government program. Technical, commercial, and potential "deep" connections among R&D projects must be identified and characterized before meaningful conclusions can be drawn about the risks being taken with taxpayers' money. Finally, proactive management of government-sponsored R&D, including interim performance reviews, are likely to yield better results than a hands-off style of management.*

Keywords: portfolios of real R&D options; R&D management; R&D project correlations; R&D risks; stage-gate reviews

Linquiti, Peter D. *The Public Sector R&D Enterprise: A New Approach to Portfolio Valuation.* New York: Palgrave Macmillan, 2015. DOI: 10.1057/9781137542090.0009.

DOI: 10.1057/9781137542090.0009

The analysis of DOE's Chemical Industries R&D program presented in the previous chapter demonstrates that it is indeed feasible to prospectively calculate in a rigorous manner the risk and potential return of a portfolio of R&D projects. Moreover, as explained below, the analysis also suggests several general principles that are relevant to the formation and valuation of public sector R&D initiatives.

First, the failure to properly measure the value of the flexibility can lead to a substantial underestimate of the value of the investment. Because R&D creates the opportunity, but not the obligation, to deploy the results of the research, and because it allows time over which evolving market conditions and technical outcomes can be observed, an R&D initiative ought not to be valued as if it were a "once and for all" decision, but should be viewed as a real option. As shown in Table 4.10, the DT/DCF method suggests that the 19 R&D projects have a net value of $504 million, while the real option method suggests that the value is over $1.34 billion, a difference of $835 million.

What's more, the switch from a DT/DCF framework to a real option approach does not have the same effect on each project. While the real option value will always be the same or higher, the degree of difference for any one project depends on the level of uncertainty and the magnitude of the upside payoff. For example, the smallest difference between the two methods occurs in the case of the Integrated Distillation project where the real option value is about 43 percent higher than the DT/DCF value. The largest difference is seen in the Oxidation Catalysts project where the real option value is higher than the DT/DCF value by a factor of more than nine. As shown in Figure 4.3, these variations suggest a different ranking of projects based on value, and hence could lead to different choices about which projects to include in a portfolio.

Accordingly, the public R&D manager interested in accurately characterizing the value of the public investment in his or her R&D program would be well-advised to avoid the DT/DCF method and instead rely on the insights that can be provided by the real option method. Doing so has the additional advantage of a more accurate ranking of the likely payoff of multiple proposed R&D projects.

Second, the analysis demonstrates the wisdom of treating the performance of a portfolio of R&D projects as more than the simple sum of the performances of the projects within it. As soon as we accept the premise that both risk and return are important, then we are compelled to conduct

DOI: 10.1057/9781137542090.0009

portfolio-level analysis. Without it, risk simply cannot be assessed. For example, the sensitivity analysis in Table 4.13 confirms Markowitz's finding that risks are not additive. In all scenarios, the expected return is identical (about $1.4 billion) while risk varies radically (ranging from a 50% chance of failing to break even to a 4 percent chance) solely as a function of the correlation among project performance. As a second example, recall that in the case of the NRC methodology (Table 4.11), it appears that there is a 32 percent chance that the 19 R&D projects will produce benefits below $67.5 million, but with a real option approach and simulation of project relationships, the risk of such an outcome drops to about 6 percent, a result that simply *cannot* be ascertained without a portfolio analysis.

Accordingly, the public R&D manager should always take a portfolio perspective on the set of investments being made. Looking only at individual R&D projects is a fundamentally unsound practice. Without a portfolio perspective, there is no way to discern how much risk is being taken with the taxpayers' money.

Third, even with a portfolio-wide approach to valuation, credible estimates of returns and risks can be obtained only after explicit incorporation in the analysis of the relationships between individual projects. To start, the interdependence of commercial results—whether the portfolio contains projects to develop competing or complementary technologies—means that portfolio returns can only be predicted on the basis of an analysis of the relationship among projects. For example, when project relationships are ignored, the predicted portfolio value is $1.367 billion, rather than $1.406 when market relationships are simulated. This is a small, but not trivial, difference. In addition, as described in Chapter 3, the opportunity to share results across projects while they are still underway, and the prospect of economies of scope across R&D projects, are two additional reasons why portfolio returns are not simply the sum of expected project returns.

As shown in Table 4.12, the simulation of project relationships also has a significant effect on risk, as measured by the variability of portfolio outcomes. While the mean portfolio value climbs from $1.367 billion to $1.406 billion when these relationships are analyzed, risks also increase. The coefficient of variability, for example, goes from goes from 1.19 to 1.41. Similarly, the risk of failing to achieve Tassey's target of a nominal 25 percent return is actually 18.2 percent while an analysis that ignores project relationships would erroneously report the risk to be 15.4 percent.

DOI: 10.1057/9781137542090.0009

When project relationships are explicitly modeled, and as in this case introduce positive correlations among projects, risk increases because the effect of diversification is muted as investments become increasingly positively correlated. While in this case, risks are higher because of the generally positive project correlations, it is also possible that, in another case, risks might be overstated if projects exhibit negative correlations.

Accordingly, the public R&D manager should always investigate whether there are potential relationships among the R&D project in his or her portfolio. While moving from project-level to portfolio-level analysis—my second suggestion above—will allow the characterization of risk, ignoring inter-project relationships will undermine the accuracy of that risk characterization. Managers should look for technical connections among projects (e.g., whether the success or failure of one project might be suggestive of the results of the other), commercial connections among projects when their outputs reach the market (e.g., whether they are substitutes or complements), and deep correlations among projects that may exist because their value is influenced by an independent, nonobvious, third variable. In this case, energy prices operate to create a deep correlation among DOE's R&D projects, but it is not hard to imagine other examples. Changes in the Medicare rules for drug reimbursement might affect the commercial value of all prescription drugs for all diseases, thus creating correlations among the values of a wide range of pharmaceutical R&D initiatives. Similarly, to the extent that climate change affects agricultural productivity, there may be deep correlations among the value of R&D projects related to different crops in different locations.

Fourth, the nonadditivity of returns creates the opportunity to use active management to increase the returns from an R&D portfolio. Public R&D managers would be well-advised to use the stage-gate review process to take advantage of interim results in a research project to make decisions about modifying, accelerating, or terminating it. Sharing facilities, personnel, and data across projects offers another means of boosting portfolio returns over the independent execution of each project. Equally important is the value of sharing interim results across projects, so that the knowledge gained in one project can be used to increase the odds of success in a second project that is taking place at the same time. The potential for sharing knowledge across project also has implications for the sequencing of R&D projects. In some cases, it may be prudent to delay a project, pending the results of another project

DOI: 10.1057/9781137542090.0009

which, if successful, could improve the prospects for the project that has been delayed.[1] The discussion in Sections 2.4 (on R&D activities) and 2.5 (on the institutional conditions that affect R&D initiatives) address this issue in greater detail.

The portfolio valuation presented in Chapter 4 makes clear that the application of these four principles is not a trivial exercise. The data needs are substantial and the calculations can be complex. These challenges are, however, an inescapable feature of government-sponsored R&D. Zvi Griliches, one of the first scholars working in the field, observed back in 1958 that:

> Conceptually, the decisions made by [a public] administrator of research funds are among the most difficult economic decisions to make and evaluate, but basically they are not very different from any other type of entrepreneurial decision. (p. 431)

Although it may seem to be a daunting task to subject major public R&D initiatives to these sorts of rigorous analyses, I believe that we have no choice but to try. Both the size of the investment—$30 billion per year in US nondefense applied R&D—and the nature of the potential payoffs—enhanced economic prosperity and solutions to profound societal problems—almost certainly justify the time and effort it takes to conduct such analyses.

Note

1   Some of these suggestions may be hard to implement if the R&D projects are being executed by different performers who are resistant to share research findings across organizational lines, or if the funding mechanism is a grant instrument that does not allow the funder any latitude to insist on a change of direction or scope. That notwithstanding, integrated management of a set of R&D projects is likely to increase the total returns from those projects.

DOI: 10.1057/9781137542090.0009

6

Next Steps in Valuing R&D Portfolios: A Research Agenda

Abstract: *Chapter 6 lays out a research agenda for continued investigation of the risks and returns of government R&D programs. Ways of incorporating social benefits, like reduced pollution, need to be further developed. Methods for better simulating the option exercise decisions of multiple firms to deploy a new technology at different points in time are also needed. In addition, more work is warranted to improve the characterization and simulation of the multiple types of potential relationships among the projects in an R&D portfolio. Finally, for R&D programs that apply the stage-gate review process, the method used earlier in the book requires an adjustment to accommodate sequential decision-making.*

Keywords: R&D management; R&D project correlations; social benefits of R&D; stage-gate reviews; technology diffusion

Linquiti, Peter D. *The Public Sector R&D Enterprise: A New Approach to Portfolio Valuation.* New York: Palgrave Macmillan, 2015. DOI: 10.1057/9781137542090.0010.

The application of real option techniques to a portfolio of R&D projects can yield considerable insight about its risks and potential returns. That said, some aspects of the methodology used in Chapter 4 could be developed further. This concluding chapter poses four questions, sound answers to which could facilitate the application of these methods to public sector R&D. For each question, I briefly sketch out a potential methodological path forward.

How should social costs and benefits be incorporated into the analysis?

As discussed in Section 2.2.1, one rationale for government support of R&D is a concern about the environmental externalities of existing technologies that act to decrease the returns to innovations in technologies to reduce those environmental impacts. DOE also aims to address other market failures:

> Public programs often have social benefits that are not valued by markets.... For the DOE [research and development] programs, two broad classes of benefits have this characteristic: the environmental benefits of energy technology and the security benefits of energy savings or energy alternatives. (NRC, 2007, p. 3)

NRC did not monetize the social benefits of reduced environmental emissions or of enhanced national security. Accordingly, it was not possible for me to incorporate such market failures in the analysis; both the option exercise decision and the resulting benefit stream were assessed from only a private perspective.

While it might be difficult to monetize the value of enhanced national security, the same is not true of environmental benefits. Recent work for the Department of Energy, in fact, retrospectively estimated the emission reductions associated with DOE-sponsored R&D in the renewable energy area (Gallaher, Rogozhin, & Petrusa, 2010). Given the reduction in emissions, human health benefits were estimated and then monetized using models developed by the Environmental Protection Agency. There is no reason why this methodology couldn't be integrated with a prospective analysis based on a real option methodology.

When it comes to adjusting the option valuation method to address these external social costs and benefits, the answer at first seems straightforward. One could add to the private energy cost savings the additional social costs and benefits of implementing the research results, but this approach is problematic. It ignores the fact that the private decision-maker would almost certainly not take account of these external costs and benefits in its decision to exercise the option.

Alternatively, an analyst could develop estimates of the social costs and benefits of the new technology and then treat option exercise as a private choice made by firms that consider only private cost and benefits. Once the private exercise decision had been simulated, the analyst would then tally the additional social costs and benefits in order to form a complete picture of net social benefits.

In government R&D, who decides to exercise the option?

Most of the scholarship on real options has focused on the individual firm and how it should manage its real options. Most government R&D programs, however, generate new knowledge that can be implemented by multiple firms. Accordingly, there is not a single option holder making the choice to exercise the option and deploy the fruits of the R&D project. Instead, as described in Section 2.7.1, multiple firms that might extract value from the research will *each* make a decision about whether to invest in converting the research into a change to its own operations, based on its assessment of benefits, costs, and uncertainties. Costs of implementation (i.e., the option exercise price) might vary among firms, as could the cost of capital or other hurdle rate for investment decisions. The timing of option exercise may also differ across firms; one might apply the research results immediately while another might leave the results "on the shelf" for a few years before incurring costs and implementing the change.

The present study does not address this issue, instead all firms are assumed to act in a homogenous fashion, making the option exercise decision at the same time based on the same calculus. Improving the realism of this aspect of the analysis is, however, largely an empirical, rather than a theoretical, challenge. With sufficient data about the potential beneficiaries of the newly developed technology, along with the cost and performance characteristics of the new technology, an analyst could simulate the option exercise decision of each firm (or group of similar firms) that would consider adopting the technology. Once the option process had been conducted at this "micro-level," the analyst would aggregate the decisions of the individual actors to assess total societal costs and benefits.

DOI: 10.1057/9781137542090.0010

How can project relationships be effectively characterized?

When it comes to characterizing the technical and market factors that potentially link one R&D project to another, expert judgment is almost certainly the best place to start. The work of the original NRC panel and of the professional chemical engineer engaged for this study suggests that such analyses are feasible. One could imagine a more sophisticated approach that created a "roadmap" to characterize the technology and market events—both inside and outside the R&D program—that could play out over time and affect the value of the technologies under development. Such a roadmap could capture network effects and the complementarity and substitutability of these technologies. A roadmap could also link the components of larger technical systems and capture the operation of an exogenous variable (such as energy prices) that create deep correlations among projects.

An expert-generated roadmap would not, however, by itself provide enough information to meaningfully characterize the risk and return of an R&D portfolio. The roadmap would need to be rendered in mathematical form, using the language of statistics, to facilitate its simulation. For this present analysis, technical interdependencies (but not commercial interdependencies) were simulated in a relatively simple fashion, using correlation coefficients, rather than more complex (and potentially more realistic) interdependence structures. Fortunately, the state of the art in risk analysis has gone well beyond simple correlation structures and now includes a wide range of copulas that can be applied to more realistically describe the behavior of related variables. A copula is simply a function that can be used to link the probabilistic distributions of two variables in a joint distribution that preserves the characteristics of the underlying univariate distribution (Clemen & Reilly, 1999; Wang & Dyer, 2012). In addition, introduction of more detailed decision tree logic (and programming code) in the Monte Carlo simulation can also capture more complex project relationships.

Should R&D projects be framed as compound, rather than simple, options?

The analysis here has focused on simple call options that expire on a date certain and are either exercised or allowed to expire unexercised. Implicit in this approach is an assumption that R&D comprises a single stage, rather than a series of sequential steps each of which ends with a decision to continue or terminate the research. As explained in Section 3.3.2,

DOI: 10.1057/9781137542090.0010

corporate R&D initiatives often are managed with a series of stage-gate reviews that provide a disciplined process for weeding out R&D projects that are underperforming expectations and to increase funding for the most promising R&D projects. Some, though certainly not all, government R&D projects are also managed with the stage-gate process. I take as a given that doing so is a sound management practice.

Real option techniques can be readily adjusted to address multistage R&D efforts by applying compound, rather than simple, option valuation. A two-stage compound option is essentially an option on an option, and an n-stage compound option is a series of options with n opportunities to exercise the option or discontinue the investment. In the context of multistage R&D, for example, the first stage of research creates an option, not on the finished technology, but on the opportunity to launch the second stage of the research (which would not have been possible without completion of the first stage). While simulating compound options is computationally more complex, and may require additional expert judgments about probabilities of technical success at the end of each stage, doing so is theoretically no different from the methodology that was used in Chapter 4.

DOI: 10.1057/9781137542090.0010

References

AAAS. (2013). *AAAS Report XXXVIII Research and Development FY2014*. Washington, DC: American Association for the Advancement of Science.

Aberman, J. (2014). *Building a Bigger Tent for Technology Innovators: The Government Is More Creative than You Think*. McLean, VA: Tandem - NSI.

Abramovitz, M. (1956). Resource and Output Trends in the United States since 1870. *The American Economic Review, 46*(2), 5–23.

Aghion, P., David, P., & Foray, D. (2009). Science, Technology, and Innovation for Economic Growth: Linking Policy Research and Practice in "STIG Systems." *Research Policy, 38*, 681–693.

Anand, J., Oriani, R., & Vassolo, R. (2007). Managing a Portfolio of Real Options. *Real Options Theory: Advances in Strategic Management, 24*, 275–303.

Arnold, E. (2004). Evaluating Research and Innovation Policy: A Systems World Needs Systems Evaluations. *Research Evaluation, 13*(1), 3–17.

ARPA-E. (2013). *ARPA-E Strategic Vision*. Advanced Research Project Agency—Energy. Washington, DC: U.S. Department of Energy.

Arrow, K. (1962). Economic Welfare and the Allocation of Resources for Invention. In R. Nelson, *The Rate and Direction of Inventive Activity* (p. 609). Princeton, NJ: Princeton University Press.

Atkinson, R., & Ezell, S. (2012). *Innovation Economics: The Race for Global Advantage*. New Haven, CT: Yale University Press.

DOI: 10.1057/9781137542090.0011

Bernanke, B. (2011). Promoting Research and Development: The Government's Role. *Issues in Science and Technology, 27*(4), 37–41.

Bernstein, P. (1998). *Against the Gods: The Remarkable Story of Risk*. New York, NY: Wiley and Sons.

Blau, G., Pekny, J., Varma, V., & Bunch, P. (2004). Managing a Portfolio of Interdependent New Product Candidates in the Pharmaceutical Industry. *Journal of Product Innovation Management, 21*(4), 227–245.

Bodie, Z., Kane, A., & Marcus, A. (2009). *Investments* (8th ed.). New York, NY: McGraw Hill.

Bodner, D., & Rouse, W. (2007). Understanding R&D Value Creation with Organizational Simulation. *Systems Engineering, 10*(1), 64–82.

Bonvillian, W. (2011). The Problem of Political Design in Federal Innovations Organization. In K. Husbands Fealng, J. Lane, J. Marburger, & S. Shipp, *The Science of Science Policy: A Handbook* (pp. 302–326). Stanford, CA: Stanford University Press.

Bonvillian, W., & Van Atta, R. (2011). ARPA-E and DARPA: Applying the DARPA Model to Energy Innovation. *Journal of Technology Transfer, 36*(5), 469–513.

Bozeman, B. (2000). Technology Transfer and Public Policy: A Review of Research and Theory. *Research Policy, 29*, 627–655.

Bozeman, B., & Rogers, J. (2001). Strategic Management of Government-Sponsored R&D Portfolios. *Environment and Planning C: Government and Policy, 19*(3), 413–442.

Brealey, R., & Myers, S. (2003). *Princples of Corporate Finance*. New York, NY: McGraw-Hill.

Brosch, R. (2008). *Portfolios of Real Options*. Berlin, Germany: Springer-Verlag.

Brown, M., Berry, L., & Goel, R. (1991). Guidelines for Successfully Tranferring Government-Sponsored Innovations. *Research Policy, 20*, 121–143.

C2ES. (2014). *Federal Vehicle Standards*. Arlington, VA: Center for Climate and Energy Solutions.

Carayannis, E., & Grigoroudis, E. (2014). Linking Innovation, Productivity, and Competitiveness: Implications for Poicy and Practice. *Journal of Technology Transfer, 39*(2), 199–218.

Casault, S., Groen, A., & Linton, J. (2013). Examination of the Behavior of R&D Returns Using a Power Law. *Science and Public Policy, 40*(2), 219–228.

DOI: 10.1057/9781137542090.0011

Chien, C. (2002). A Portfolio-Evaluation Framework for Selecting R&D Projects. *R&D Management, 32*(4), 359–368.

Childs, P., & Triantis, A. (1999). Dynamic R&D Investment Policies. *Management Science, 45*(10), 1359–1377.

Childs, P., Ott, S., & Triantis, A. (1998). Capital Budgeting for Interrelated Projects. *Journal of Financial and Quantitative Analysis, 33*(5), 305–334.

Chow, B., Silberglitt, R., & Hiromoto, S. (2009). *Toward Affordable Systems: Portfolio Analysis and Management for Army Science and Technology Projects*. Santa Monica, CA: Rand Corporation.

Chow, B., Silberglitt, R., Hiromoto, S., Reilly, C., & Panis, C. (2011). *Toward Affordable Systems II: Portfolio Management for Army Science and Technology Programs under Uncertainties*. Santa Monica, CA: Rand Corporation.

Chu, S. (January 21, 2010). Secretary of Energy. *Statement before Senate Committee on Energy and Natural Resources*. Washington, DC.

Clemen, R., & Reilly, T. (1999). Correlations and Copulas for Decision and Risk Analysis. *Management Science, 45*(2), 208–224.

Cohen, L., & Noll, R. (1991). *The Technology Pork Barrel*. Washington, DC: Brookings Institution.

Colvin, M., & Marvelias, C. (2011). R&D Pipeline Management: Task Interdependencies and Risk Management. *European Journal of Operational Research, 215*(3), 616–628.

Congressional Budget Office. (2014). *Estimated Impact of the American Recovery and Reinvestment Act on Employment and Economic Output in 2013*. Washington, DC.

Cooper, R., Edgett, S., & Kleinschmidt, E. (2001). Portfolio Management for New Product Development: Results of an Industry Practices Study. *R&D Management, 31*(4), 361–380.

Copeland, T., & Antikarov, V. (2003). *Real Options: A Practitioner's Guide*. New York, NY: Cengage.

David, P. (1985). Clio and the Economics of QWERTY. *American Economic Review, Papers and Proceedings, 75*(2), 332–337.

Davis, G., & Owens, B. (2003). *Optimizing the Level of Renewable Electric R&D Expenditures Using Real Option Analysis*. NREL/TP-620-31221. Golden, CO: National Renewable Energy Laboratory.

de Neufville, R., Hodota, K., Sussman, J., & Scholtes, S. (2008). Real Options to Increase the Value of Intelligent Transportation Systems. *Transportation Research Record, 2086*, 40–47.

DOI: 10.1057/9781137542090.0011

deLeon, P., & Martell, C. (2008). Policy Sciences Approach. In E. Berman, *Encyclopedia of Public Administration and Public Policy* (2nd ed., pp. 1495–1498). London, UK: Informa Ltd.

Dias, M. (2006). *Real Option Theory for Real Asset Portfolios: The Oil Exploration Case.* Retrieved from www.puc-rio.br/marco.ind/pdf/dias_portfolio_ep.pdf

Dickinson, M., Thornton, A., & Graves, S. (2001). Technology Portfolio Management: Optimizing Interdependent Projects over Multiple Time Periods. *IEEE Transactions on Engineering Management, 48*(4), 518–527.

Dixit, A., & Pindyck, R. (1994). *Investment under Uncertainty.* Princeton, NJ: Princeton University Press.

Dudley, S., & Brito, J. (2012). *Regulation: A Primer* (2 ed.). Arlington, VA: Mercatus Center & George Washington University.

Edquist, C. (2005). Systems of Innovation: Perspectives and Challenges. In J. Fagerberg, D. Mowery, & R. Nelson, *Oxford Handbook of Innovation* (pp. 181–208). New York: Oxford University Press.

EIA. (2012). *Most States Have Renewable Portfolio Standards.* Energy Information Agency. Washington, DC: Department of Energy.

EIA. (January 1994). *Annual Energy Outlook 1994.* Energy Information Administration. Washington, DC: Department of Energy.

EIA. (January 1995). *Annual Energy Outlook 1995.* Energy Information Administration. Washington, DC: Department of Energy.

EIA. (January 1996). *Annual Energy Outlook 1996.* Energy Information Administration. Washington, DC: Department of Energy.

EIA. (December 1996). *Annual Energy Outlook 1997.* Energy Information Administration. Washington, DC: Department of Energy.

EIA. (December 1997). *Annual Energy Outlook 1998.* Energy Information Administration. Washington, DC: Department of Energy.

EIA. (December 1998). *Annual Energy Outlook 1999.* Energy Information Administration. Washington DC: Department of Energy.

EIA. (December 1999). *Annual Energy Outlook 2000.* Energy Information Administration. Washington, DC: Department of Energy.

EIA. (December 2000). *Annual Energy Outlook 2001.* Energy Information Administration. Washington, DC: Department of Energy.

EIA. (December 2001). *Annual Energy Outlook 2002.* Energy Information Administration. Washington, DC: Department of Energy.

EIA. (January 2003). *Annual Energy Outlook 2003.* Energy Information Administration. Washington, DC: Department of Energy.

DOI: 10.1057/9781137542090.0011

EIA. (January 2004). *Annual Energy Outlook 2004*. Energy Information Administration. Washington, DC: Department of Energy.

Eliat, H., Golany, B., & Shtub, A. (2006). Constructing and Evaluating Balanced Portfolios of R&D Projects with Interactions: A DEA Based Methodology. *European Journal of Operational Research, 172*(3), 1018–1039.

European Commission. (March 19, 2014). *Taking Stock of the Europe 2020 Strategy for Smart, Sustainable, and Inclusive Growth*. Brussels, Belgium.

Eurostat. (October 13, 2014). *R & D Expenditure*. Retrieved from ec.europa.eu/eurostat.

Evans, R., Hinds, S., & Hammock, D. (2009). Portfolio Analysis and R&D Decision Making. *Nature Reviews Drug Discovery, 8*(3), 189–190.

Executive Office of the President. (2010). *Science and Technology Priorities for the FY2012 Budget*. Washington, DC: Offices of Management & Budget and of Science and Technology Policy.

Executive Office of the President. (2014). *Science and Technology Priorities for the FY2016 Budget*. Washington, DC: Offices of Management & Budget and of Science and Technology Policy.

Executive Office of the President. (February 2011). *A Strategy for American Innovation: Securing Our Economic Growth and Prosperity*. Washington, DC: Offices of Management & Budget and of Science and Technology Policy.

Feller, I. (2011). Science of Science and Innovation Policy: The Emerging Community of Practice. In K. Husbands Fealing, J. Lane, J. Marburger, & S. Shipp, *The Science of Science Policy: A Handbook* (pp. 131–155). Stanford, CA: Stanford University Press.

Flanagan, K., Uyarra, E., & Laranja, M. (2011). Reconceptualising the "Policy Mix" for Innovation. *Research Policy, 40*, 702–713.

Fri, R. (November 20–21, 2009). *Personal Email Communications*.

Gallaher, M., Rogozhin, A., & Petrusa, J. (2010). *Retrospective Benefit-Cost Evaluation of U.S. DOE Geothermal Technologies R&D Program Investments: Impacts of a Cluster of Energy Technologies*. RTI International. Washington, DC: Dept of Energy, Office of Energy Efficiency & Renewable Energy.

GAO. (2012). *Designing Evaluations: 2012 Revision*. Applied Research and Methods Division. Washington, DC: United States Government Acountability Office.

Georghiou, L., & Roessner, D. (2000). Evaluating Technology Programs: Tools and Methods. *Research Policy, 29*, 657–678.

DOI: 10.1057/9781137542090.0011

Giebe, T., Grebe, T., & Wolfstetter. (2006). How to Allocate R&D (and Other) Subsidies: An Experimentally Tested Policy Recommendation. *Research Policy, 35*(9), 1261–1272.

Goel, R. (1999). On Contracting for Uncertain R&D. *Managerial and Decision Economics, 20*(2), 99–106.

Goel, R., & Rich, D. (2005). Organization of Markets for Science and Technology. *Journal of Institutional and Theoretical Economics, 161*(1), 1–17.

Goel, R., Payne, J., & Ram, R. (2008). R&D Expenditures and U.S. Economic Growth: A Disaggregated Approach. *Journal of Policy Modeling, 30*, 237–250.

Golabi, K., Kirkwood, C., & Sicherman, A. (1981). Selecting a Portfolio of Solar Energy Projects Using Multiattribute Preference Theory. *Management Science, 27*(2), 174–189.

Graetz, M. (2012). Energy Policy: Past or Prologue? *Daedalus, 141*(2), 31–44.

Griliches, Z. (1958). Research Costs and Social Returns: Hybrid Corn and Related Innovations. *Journal of Political Economy, 66*, 419–431.

Griliches, Z. (1992). The Search for R&D Spillovers. *Scandanavian Journal of Economics, 94*, 29–47.

Halchin, L. (2011). *Other Transaction (OT) Authority.* Washington, DC: Congressional Research Service.

Hall, B. (2005). Innovation and Diffusion. In J. Fagerberg, D. Mowery, & R. Nelson, *The Oxford Handbook of Innovation* (pp. 459–484). New York: Oxford University Press.

Hall, B., Mairesse, J., & Mohnen, P. (2009). *Measuring the Returns to R&D.* Cambridge, MA: National Bureau of Economic Research.

Holland, J. (October 29, 2009). Presentation: ERDC Portfolio Management. *Science of Science Policy Workshop.* Washington, DC: U.S. Army Corps of Engineers.

Hull, J. (2009). *Options, Futures, and Other Derivatives* (7th ed.). Upper Saddle River, NJ: Pearson Education.

Ibbotson, R., & Chen, P. (2003). Long-Run Stock Returns: Participating in the Real Economy. *Financial Analysts Journal, 59*(1), 88–98.

ITG. (2008). *The Science of Science Policy: A Federal Research Roadmap.* Office of Science & Technology Policy, Interagency Task Group. Washington, DC: Executive Office of the President.

Jaffe, A. (2011). Analysis of Public Research, Industrial R&D, and Commercial Innovation: Measurement Issues Underlying the Science

DOI: 10.1057/9781137542090.0011

of Science Policy. In K. Husbands Fealing, J. Lane, J. Marburger, & S. Shipp, *The Science of Science Policy: A Handbook* (pp. 193–207). Stanford, CA: Stanford University Press.

Jaffe, A., Newell, R., & Stavins, R. (2004). *A Tale of Two Market Failures: Technology and Environmental Policy.* RFF DP 04-38. Washington, DC: Resources for the Future.

Johnson, S. (2011). *Where Good Ideas Come From.* New York: Riverhead Trade.

Jordan, G. (2011). Presentation: The Diverse Variables to Consider When Planning and Assessing Public Research Portfolios. *Atlanta Science & Innovation Policy Conference.* Atlanta, GA: Sandia National Laboratories.

Jordan, G. (2013). Logic Modeling: A Tool for Designing Program Evaluations. In A. Link, & N. Vonortas, *Handbook on the Theory and Practice of Program Evaluation* (pp. 143–165). Northampton, MA: Edward Elgar Publishing.

Kamien, M., & Schwartz, N. (1982). *Market Structure and Innovation.* Cambridge, Enlgand: Cambridge University Press.

Kester, C. (1984). Today's Options for Tomorrow's Growth. *Harvard Business Review, 62*(3), 153–160.

Kingdon, J. (2011). *Agendas, Alternatives, and Public Policies* (2nd ed.) Glenview, IL: Longman.

Kingsley, G., Bozeman, B., & Coker, K. (1996). Technology Transfer and Absorption: An "R&D Value-Mapping" Approach to Valuation. *Research Policy, 25,* 967–995.

Lanza, R. (2012). *Characterizing the Relationships among a Set of Research and Development Projects Sponsored by the U.S. Department of Energy to Improve Energy Efficiency in the Chemicals Industry.* Washington, DC: ICF Technology, LLC.

Link, A., & Scott, J. (2011). *Public Goods, Public Gains: Calculating the Social Benefits of Public R&D.* New York: Oxford University Press.

Link, A., & Scott, J. (2012). *The Theory and Practice of Public-Sector R&D Economic Impact Analysis.* Planning Report 11-1. Gaithersburg, MD: National Institute of Standards and Technology.

Link, A., & Scott, J. (2013). *Public R&D Subsidies, Outside Private Support, and Employment Growth.* Greensboro, NC: University of North Carolina Department of Economics.

Link, A., & Vonortas, N. (2013). *Handbook on the Theory and Practice of Program Evaluation.* Northampton, MA: Edward Elgar Publishing.

DOI: 10.1057/9781137542090.0011

Linquiti, P. (2012a). The Importance of Integrating Risk in Retrospective Evaluations of Research and Development. *Research Evaluation, 21*(2), 152–165.

Linquiti, P. (2012b). *Application of Finance Theory and Real Option Techniques to Public Sector Investments Made under Uncertainty— Research & Development and Climate Change.* George Washington University: Doctoral Dissertation.

Linton, J., Morabito, J., & Yeomans, J. (2007). An Extension to a DEA Support System Used for Assessing R&D Projects. *R&D Management, 37*(1), 29–36.

Linton, J., Walsh, S., & Morabito, J. (2002). Analysis, Ranking, and Selection of R&D Projects in a Portfolio. *R&D Management, 32*(2), 139–148.

Litvinchev, I., Lopez, F., Alvarez, A., & Fernandez, E. (2010). Large-Scale Public R&D Portfolio Selection by Maximizing a Biobjective Impact Measure. *IEEE Transactions on Systems, Man, and Cybernetics, 38*(3), 572–582.

MacMillan, I., & McGrath, R. (2004). Crafting R&D Project Portfolios. In M. Tushman, & P. Anderson, *Managing Strategic Innovation and Change* (pp. 347–359). New York, NY: Oxford University Press.

Mahnovski, S. (2007). *Robust Decisions and Deep Uncertainty: An Application of Real Options to Public & Private Investment in Hydrogen and Fuel Cell Technologies.* Dissertation. Santa Monica, CA: Pardee Rand Graduate School.

Malerba, F. (2002). Sectoral Systems of Innovation and Production. *Research Policy, 31*(2), 247–264.

Mansfield, E. (1961). Technical Change and the Rate of Imitation. *Econometrica, 29*(4), 741–766.

Markowitz, H. (1952). Portfolio Selection. *Journal of Finance, 7*(1), 77–91.

MobiLens. (March 7, 2014). Retrieved November 2, 2014, from www.iclarified.com/38928/apple-keeps-gaining-us-smartphone-share-charts.

Moniz, E. (2012). Stimulating Energy Technology Innovation. *Daedalus, 141*(2), 81–93.

Myers, S. (1977). Determinants of Corporate Borrowing. *Journal of Financial Econoics, 5*(2), 147–175.

Myers, S. (1984). Finance Theory and Financial Strategy. *Interfaces, 14*(1), 126–137.

National Academy of Science. (2007). *Rising above the Gathering Storm: Energizing and Employing American for a Brighter Economic Future.* Washinton, DC: National Academies Press.

DOI: 10.1057/9781137542090.0011

National Science Foundation. (2012). *Federal Funds for Research and Development: Fiscal Years 2009–11.* Arlington, VA.

National Science Foundation. (2013). *InfoBrief 13-317: Federal Research and Development and R&D Plant Obligations Show Modest Growth in FY2010.* Arlington, VA.

National Science Foundation. (2014a). *Master Government List of Federally Funded R&D Centers.* Arlington, VA.

National Science Foundation. (2014b). *Science and Engineering Indicators.* Arlington, VA.

National Science Foundation. (December 2013). *National Patterns of R&D Resources: 2011–12 Data Update.* Arlington, VA.

Nelson, R. (1959). The Simple Economics of Basic Scientific Research. *Journal of Political Economy, 67*(3), 297–306.

Nelson, R., & Rosenberg, N. (1993). Technical Innovation and National Systems. In R. Nelson, *National Innovation Systems: A Comparative Analysis* (pp. 3–22). New York: Oxford University Press.

Newcomer, K., Hatry, H., & Wholey, J. (2010). Planning and Designing Useful Evaluations. In J. Wholey, H. Hatry, & K. Newcomer, *Handbook of Practical Program Evaluation* (3rd ed., pp. 5–29). San Francisco, CA: John Wiley & Sons.

NRC. (2001). *Energy Research at DOE: Was It Worth It?* Washington, DC: National Academies Press.

NRC. (2005). *Prospective Evaluation of Applied Energy Research and Development at DOE (Phase One): A First Look Forward.* Washington, DC: National Academies Press.

NRC. (2007). *Prospective Evaluation of Applied Energy Research and Development at DOE (Phase Two).* Washington, DC: National Academies Press.

NRC. (2013). *Sustainability for the Nation: Resource Connections and Governance Linkages.* Washington, DC: National Academies Press.

OECD. (2014a). *OECD Science, Technology and Industry Outlook 2014: Highlights.* Paris, France: OECD Publishing.

OECD. (2014b). *OECD Science, Technology and Industry Outlook 2014.* Paris, France: OECD Publishing.

Ozokweleu, D. (January 8, 2010). *Personal Email Communication.*

Peerenboom, J., Buehring, W., & Joseph, T. (1989). Selecting a Portfolio of Environmental Programs for a Synthetic Fuels Facility. *Operations Research, 37*(5), 689–699.

DOI: 10.1057/9781137542090.0011

Powell, J. (2006). *Toward a Standard Benefit-Cost Methodology for Publicly Funded Science and Technology Programs.* Gaithersburg, MD: National Institute of Standards and Technology.

Preble, C. (2003). Who Ever Believed in a Missile Gap: John F. Kennedy and the Politics of National Security. *Presidential Studies Quarterly, 33*(4), 801–826.

Pyper, J. (September 5, 2014). Elon Musk's Tesla Picks Nevada to Host Battery Gigafactory. *Scientific American.*

Ragsdale, C. (2008). *Spreadsheet Modeling and Decision Analysis* (5th ed.). Mason, OH: Thomson South-Western.

Ram, R., & Goel, R. (2009). Irreversible Investments: A Cost Benefit Perspective. In R. Brent, *Handbook of Research on Cost-Benefit Analysis* (pp. 455–481). Cheltenham, England: Edwrd Elgar.

Reinganum, J. (1989). The Timing of Innovation: Research, Development, and Diffusion. In R. Schmalensee, & R. Willig, *Handbook of Industrial Organization, Volume 1* (pp. 850–908). Elsevier Science Publishers.

Rogers. (1995). *Diffusion of Innovations* (4th ed.). New York: Free Press.

Rogers, M., Gupta, A., & Maranas, C. (2002). Real Options Based Analysis of Optimal Pharmaceutical Research and Development Portfolios. *Industrial & Engineering Chemistry Research, 41*(25), 6607–6620.

Rosenberg, N. (1972). Factors Affecting the Diffusion of Technology. *Explorations in Economic History, 10*(1), 3–33.

Ruegg, R., & Feller, I. (2003). *A Toolkit for Evaluating Public R&D Investment: Models, Methods, and Findings from ATP's First Decade.* Gaithersburg, MD: National Institute of Standards and Technology.

Ruegg, R., & Jordan, G. (2010). New Benefit-Cost Studies of Renewable and Energy Efficiency Programs of the U.S. Department of Energy: Methodology and Findings. *International Energy Program Evaluation Conference.* Paris, France.

Ruttan, V. (2006). Will Government Programs Spur the Next Breakthrough? *Issues in Science and Technology, 27*(2), 55–61.

Sapolsky, H., & Taylor, M. (2011). Politics and the Science of Science Policy. In K. Husbands Fealing, J. Lane, J. Marburger, & S. Shipp, *The Science of Science Policy: A Handbook* (pp. 31–55). Stanford, CA: Stanford University Press.

Sargent, J. (2014). *Federal Research and Development Funding: FY2015.* Washington, DC: Congressional Research Service.

DOI: 10.1057/9781137542090.0011

Schacht, W. (2005). *The Advanced Technology Program*. Washington, DC: Congressional Research Service.

Schacht, W. (2013). *The National Institute of Standards and Technology: An Appropriations Overview*. Washington, DC: Congressional Research Service.

Schilling, M. (2010). *Strategic Management of Technological Innovation* (3rd ed.). New York, NY: McGraw-Hill/Irwin.

Schooley, J. (2000). *Responding to National Needs*. Washington, DC: U.S. Government Printing Office.

Schumpeter, J. (1943). *Capitalism, Socialism, and Democracy*. London, United Kingdom: G. Allen & Unwin.

Shockley, R. (2007). *An Applied Course in Real Options Valuation*. Mason, OH: Thomson Higher Education.

Sick, G., & Gamba, A. (2005). *Some Important Issues Involving Real Options: An Overview*. Working Paper.

Sidebottom, D. (2003). Intellectual Property in Federal Government Contracts: The Past, the Present, and One Possible Future. *Public Contract Law Journal, 33*(1), 63–97.

Silberglitt, R., Sherry, L., Wong, C., Tseng, M., Ettedgui, E., Watts, A., & Stothard, G. (2004). *Portfolio Analysis and Management for Naval Research and Development*. Santa Monica, CA: Rand Corporation.

Smith, J., & Thompson, R. (2009). Rational Plunging and the Option Value of Sequential Investment: The Case of Petroleum Exploration. *The Quarterly Review of Economics and Finance, 49*(3), 1009–1033.

Smith, K. (2000). Innovation as a Systemic Phenomenon: Rethinking the Role of Policy. *Enterprise & Innovation Management Studies, 1*(1), 73–102.

Sokolov, D. (April 26, 2012). Minority Staff Director, Subcommittee on Research and Science Education, Committee on Science, Space, and Technology, U.S. House of Representatives. *Remarks at 37th Annual AAAS Forum on Science and Technology Policy*. Washington, DC.

Solak, S., Clarke, J., Johnson, E., & Barnes, E. (2010). Optimization of R&D Project Portfolios under Endogeneous Uncertainty. *European Journal of Operational Research, 207*(1), 420–433.

Solow, R. (1957). Technical Change and the Aggregate Production Function. *The Review of Economics and Statistics, 39*(3), 312–320.

St. John, J. (February 27, 2014). ARPA-E's Success Stories Gained $625M in Private Sector Buy-In. *GreenTech Media*.

DOI: 10.1057/9781137542090.0011

Stone, D. (2012). *Policy Paradox: The Art of Political Decision Making* (3rd ed.). New York, NY: Norton & Company.

Tassey, G. (1997). *The Economics of R&D Policy*. Westport, CT: Quorum Books.

Tassey, G. (2003). *Methods for Assessing the Economic Impacts of Government R&D*. Gaithersburg, MD: National Institute for Standards and Technology.

Tassey, G. (2005). Underinvestment in Public Good Technologies. *Journal of Technology Transfer, 30*(1–2), 89–113.

The Economist. (June 8, 2013). Energy: Blown Away.

Triantis, A. (2001). Real Options: State of the Practice. *Journal of Applied Corporate Finance, 14*(2), 8–24.

Triantis, A. (2003). Real Options. In D. Logue, & J. Seward, *Handbook of Modern Finance* (pp. D1–D32). New York, NY: Research Institute of America.

Triantis, A. (Forthcoming). Real Options. In P. Moles, *Encyclopedia of Financial Engineering and Risk Management*. London, United Kingdom: Fitzroy Dearborn.

Trigeorgis, L. (1996). *Real Options: Managerial Flexibility and Strategy in Resource Allocation*. Cambridge, MA: MIT Press.

U.S. Congress. (1992). Public Law 102-245. *American Technology Preeminence Act of 1991*. Washington, DC.

U.S. Congress. (1999). *House Report 106-479*. Washington, DC: Government Printing Office.

U.S. Congress. (2002). *House Report 107-564*. Washington, DC: Government Printing Office.

U.S. Congress. (2007a). *House Report 110-185*. Washington, DC: Government Printing Office.

U.S. Congress. (2007b). Public Law 110-69. *America COMPETES Act*. Washington, DC.

U.S. EPA. (2012). *Gas Guzzler Tax*. Retrieved November 2, 2014, from www.epa.gov/fueleconomy/guzzler.

U.S. Office of Management and Budget. (2003). *Circular A-4: Regulatory Analysis*.

U.S. Office of Management and Budget. (2011). *Economic Report of the President*. Washington, DC: Government Printing Office.

van Bekkum, S., Pennings, E., & Smit, H. (September 2009). A Real Options Perspective on R&D Portfolio Diversification. *Research Policy, 38*(7), 1150–1158.

DOI: 10.1057/9781137542090.0011

Vonortas, N. (2013). Social Network Methodology. In A. Link, & N. Vonortas, *Handbook on the Theory and Practice of Program Evaluation* (pp. 193–246). Northampton, MA: Edward Elgar Publishing.

Vonortas, N., & Desai, C. (2007). "Real Options" Framework to Assess Public Research Investments. *Science and Public Policy, 34*(10), 699–708.

Vonortas, N., & Hertzfeld, H. (1998). Research and Development Project Selection in the Public Sector. *Journal of Public Policy Analysis and Management, 17*(4), 621–638.

Wagner, H. (1975). *Principles of Operations Research* (2nd ed.). Englewood Cliffs, NJ: Prentice-Hall.

Wang, T., & Dyer, J. (2012). A Copulas-Based Approach to Modeling Dependence in Decision Trees. *Operations Research, 60*(1), 225–242.

Wholey, J., Hatry, H., & Newcomer, K. (2010). *Handbook of Practical Program Evaluation* (3rd ed.). San Francisco, CA: John Wiley & Sons.

Wouters, M., Roorda, B., & Gal, R. (2011). Managing Uncertainty During R&D Projects: A Case Study. *Research Technology Management, 54*(2), 37–46.

Zuniga-Vicente, J., Alonso-Borrego, C., Forcadell, F., & Galan, J. (2014). Assessing the Effect of Public Subsidies on Firm R&D Investment: A Survey. *Journal of Economic Surveys, 28*(1), 36–67.

DOI: 10.1057/9781137542090.0011

Index

DOI: 10.1057/9781137542090.0012

DOI: 10.1057/9781137542090.0012